BUILDING with LOGS

FACSIMILE REPRODUCTION 1973

THE SHOREY BOOK STORE
815 Third Avenue
Seattle, Washington 98104

PRINTED IN U.S.A.

SJU # 30

FOURTH PRINTING
Of Facsimile Reprint

Limited to 5000 Copies
June 1974

The art of log construction is relatively simple, once a few basic principles are understood. The pioneers who opened the lands beyond the eastern seaboard did not have boards with which to build such shelter as they needed. Logs were so plentiful in the forested area of our country that, with their resourceful ingenuity, the settlers built their homes in conformity with those principles of log construction which prevailed in the countries from which they migrated. Those principles have remained the same down through the years.

The pioneer had but an ax for a tool and consequently made only those articles which could be hewed out of wood. Today there are many tools available, and to do a first class job of log construction one must know how to handle the double-bitted or single-bitted ax, the broadax, saw, adz, chisel, slick, ship auger, and drawknife. In this bulletin it is assumed that the reader is familiar with the ordinary frame building methods used where wood is the principal construction material.

Washington, D. C. Issued September 1945

ISBN #0-8466-6030-X
(In Paper Covers)

BUILDING WITH LOGS

By Clyde P. Fickes, *Engineer*, and W. Ellis Groben, *Chief Architect, Forest Service*

Contents

BUILDING THE FOUNDATION

A building should have a good foundation, and a log structure is no exception to the rule. For the sake of economy in labor and material it is sufficient, in some instances, to place small buildings on piers of concrete or rough native stone, but usually it will be more satisfactory to use continuous walls of stone masonry or concrete to provide uninterrupted support for the logs and thus avoid their tendency to sag. These walls, however, should be provided with small openings for the circulation of air to prevent the wood from dry rotting. Furthermore, the continuous foundation wall has the additional advantage of preventing rodents from getting under the building. In no case should the logs be placed directly upon the ground since wood tends to decay when in contact with the earth.

The two end walls of the exterior foundation should be higher than the side walls in order to offset the difference in level of the logs on adjacent walls, the end-wall logs being half their thickness higher than those on the side walls.

In building a log wall the chief problem is in closing the opening between each pair of logs. There are various ways of doing this, but only those regarded as most satisfactory will be described in this publication. The width of such openings is affected by several factors: (1) The manner of placing the logs upon each other; (2) the type of corner used where two walls meet; (3) the openings for doors and windows; and (4) the natural shrinkage of wood in the process of drying.

PREPARING THE LOGS

The selection of straight, smooth, even-sized logs is the prime consideration (fig. 1). Top diameters should be as uniform as possible, but as a rule not less than 10 nor more than 12 inches. (Slightly

FIGURE 1.—Starting to build the log cabin—laying the foundation.

smaller or larger dimensions may be used if no others are available.) The taper should be as slight as possible. For logs longer than 40 feet, the top diameter may be less than 10 inches in order to avoid an excessive diameter at the large or butt end.

Cedar, pine, fir, and larch, in the order named, are most desirable for log construction. All knots, limbs, or bumps should be trimmed off carefully when the log is peeled. It is best to cut the logs in late fall or winter, for two important reasons: (1) Logs cut in spring or summer peel easier, but crack or check to an undesirable degree while seasoning. (2) Insect activity is dormant during the winter months; hence, if the logs are cut and seasoned then, they are less liable to damage by insects or rot-producing fungi.

Logs should be cut, peeled, and laid on skids well above the ground for at least 6 months before being placed in the building. This may not always be possible, but it is a good rule to follow. Logs should be stored in a single deck with 2 or 3 inches between them to permit complete exposure to the air. Logs having a sweep or curve should be piled with the curve uppermost so that their weight will tend to straighten them while they are drying. Where the skidding space is limited, logs may be double-decked, using poles between tiers. Unrestricted air circulation materially aids seasoning.

Sort the logs carefully before starting construction, using the better ones in the front or other conspicuous walls of the building. If the logs are not uniform in size, the larger ones should be placed at the bottom of the walls.

DIMENSIONS OF THE BUILDING

For practical reasons the dimensions of a log building are the inside measurements taken from one log to the corresponding log in the opposite wall. Outside dimensions vary somewhat with the size of the logs, thus accounting for the use of inside measurements.

Where projecting corners are desired, logs should be at least 6 feet longer than the inside dimensions of the building. In erecting the walls, the logs should be kept even or plumb on the inside faces if it is desired to finish the interior with wallboard or plaster.

FRAMING THE CORNERS

The corner is one of the most important aspects of log construction. On it the appearance and stability of the structure depend. Different types of corner construction are in use in the United States, each varying in accordance with local building customs or individual taste.

FIGURE 2.—The round-notch or saddle corner. This is an unusually fine example of scribing and fitting logs together. The square-cut logs have yet to be dressed and shaped with the ax to give them a pleasing appearance.

Round-Notch Corner

The round-notch, or saddle, corner (fig. 2) is generally considered the most satisfactory from every standpoint. This type of corner gives the most distinctive appearance because the logs project suffi-

FIGURE 3.—Ranger station, Gallatin National Forest, Mont., illustrating effective use of round-notch corners. A, and B, Dwelling under construction; C, barn.

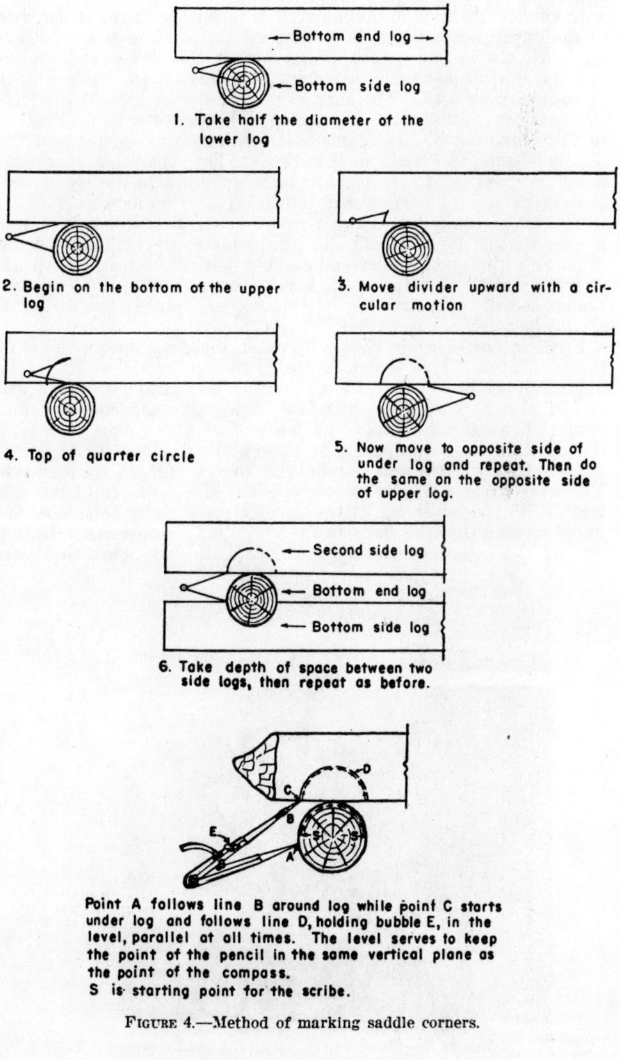

FIGURE 4.—Method of marking saddle corners.

ciently beyond the corner not to appear dubbed off (fig. 3). It is a good, self-locking, mechanical joint, relatively easy to construct, and holds the logs rigidly in place.

In cutting the saddle, the material is taken out of the under side of the upper log without disturbing the top surface of the bottom log. All the moisture thus drains out at the corner and, consequently, the wood is much less subject to decay than if other types of corners were used. The shrinkage in the outer area of the log's circumference tends to open up the space between the logs. Finally, in the round-notch corner, one-half of the shrinkage between the logs is allowed to remain in the corner. The separation, therefore, is not as great as if each log had been cut down to the heartwood, a disadvantage common to most other types of corners.

The tools required to make a round-notch or saddle corner are: A pair of log dogs to hold the log in place, 10- or 12-inch wing dividers with pencil holder and level-bubble attachment, sharp ax, 2-inch gouge chisel with outside bevel, crosscut saw, spirit level, and plumb board. The framing of this corner, described in figure 4, should be relatively easy.

First, the bottom logs should be set in place on opposite sides of the foundation. Hew a flat face of 2 to 3 inches in width on the under side of the log where it rests on the foundation, so that it will lay in place. Then place the bottom log on each end wall and accurately center it so that the inside face of all four logs is to the exact interior dimensions of the building. Dog the logs into place so they will not move while being marked for the corner notch. The wing divider is now set for one-half the diameter of the side log. With the lower leg of the divider resting on the side of the under log and the other leg, with the level bubble uppermost, resting against the bottom of the upper log and directly above the lower

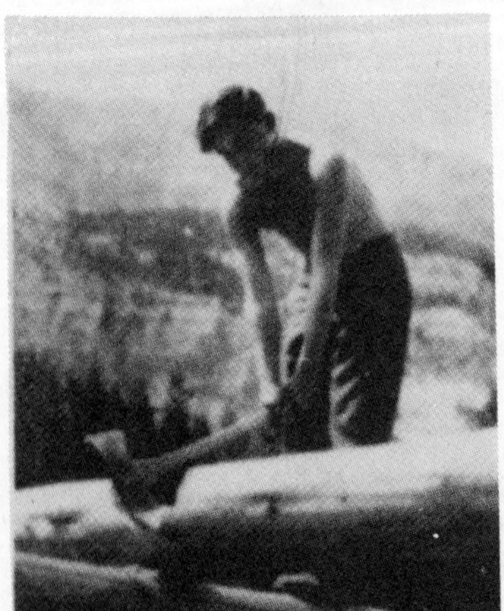

FIGURE 5.—Chopping the notch in a saddle corner.

log, start moving the divider upward, with a side motion, so that the lower leg follows the curvature of the under log. The pencil point of the upper leg makes a mark on the surface of the upper log which will be the intersection of the surfaces of the two logs when the notch has been cut from the upper one. Repeat this operation four times to mark all four sides of the corner. A little practice will make you adept at keeping the points of the divider perpendicular to each other.

After the notch has been marked at both ends of the log, turn it over on its back. It is a good idea to intensify the divider mark with an indelible pencil so that it will be easily followed. Chop the notch out roughly, as illustrated in figure 5, then chip down as closely as possible to the mark, supplying the finishing touches with a gouge chisel. The finished notch should be cupped out just enough to allow the weight of the log to come on the outside edges, thus insuring a tight joint.

When the next side log is rolled into place, the dividers should be set apart for the width of the space between the top of the first and the bottom of the following log, and the marking repeated as before. If you wish to have the upper log "ride" the lower one a little, so that an especially tight joint is obtained, the dividers should be set a little wider apart than the space actually requires.

Other Log Corners

The dovetail, or box, corner (figs. 6 and 7) is a strong corner, and considerable experience is required in order to make a neat-looking job. This type has several undesirable features: (1) The logs are apt to develop a wide crack because the corner is framed from the part of the log in which the least shrinkage occurs, and (2) since the logs are hewed down to form the corner, the wood has a tendency to collect and retain moisture which soon results in decay. Also, this corner detracts noticeably from the "loggy" appearance so characteristic and desirable in log structures. The drawings in figure 6 show the most practical methods of marking and framing the dovetail, or box, corner.

The flat, or plain, tenon corner (fig. 8), is also common. It may be made in two ways. In one, only the bearing surfaces are framed, while in the other, all four sides of the tenon are framed flat. The plain tenon corner does not have the highly desirable feature of being self-locking. However, it is simple to make and economical, and therefore especially suitable for temporary structures. The logs must be pinned together, as shown in figure 11. All the framing can be done on the ground, before the logs are put in place. Carefully fitted, this makes a neat-looking job.

Directions for constructing the flat, or plain, tenon corner.—Square one end of log, as in figure 8, at point *A*, then measure required length and saw the opposite end square, at *B*. If the log has any curvature, turn it on the skids until its back is up. Determine the thickness of the tenons, based upon the average top and butt diameters of the log. Then take an 18-inch length of board the same width as the thickness of the tenons, driving a nail through its center and into the center of the log. Place the spirit level on top of the board and mark lines on the log at the top and bottom

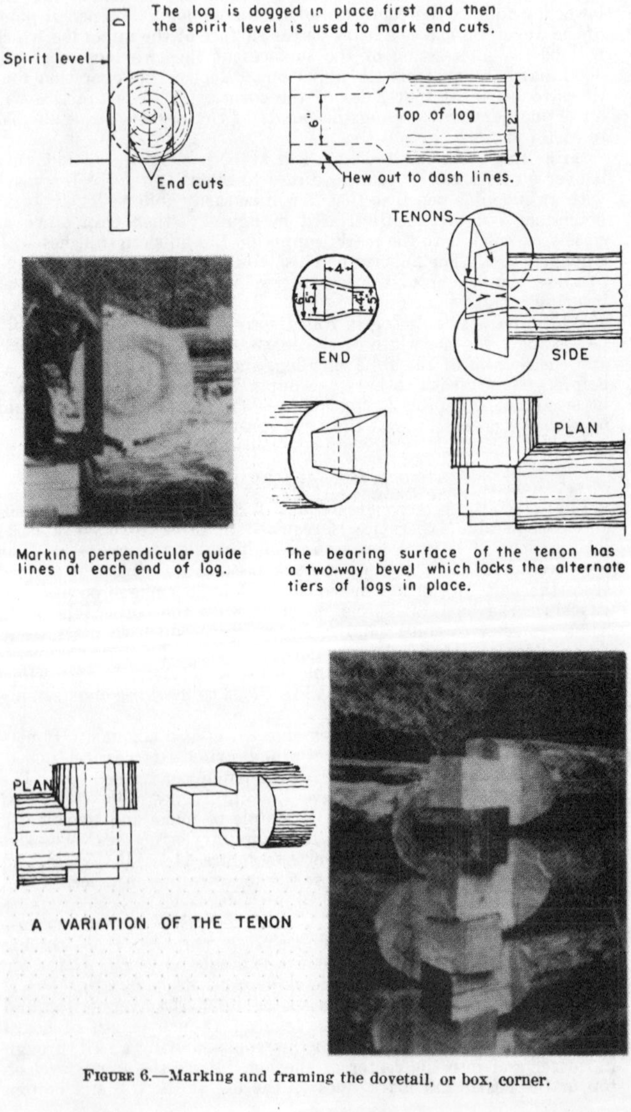

FIGURE 6.—Marking and framing the dovetail, or box, corner.

FIGURE 7.—Ranger Station, Lolo National Forest, Mont. Note the meticulous construction of box corners.

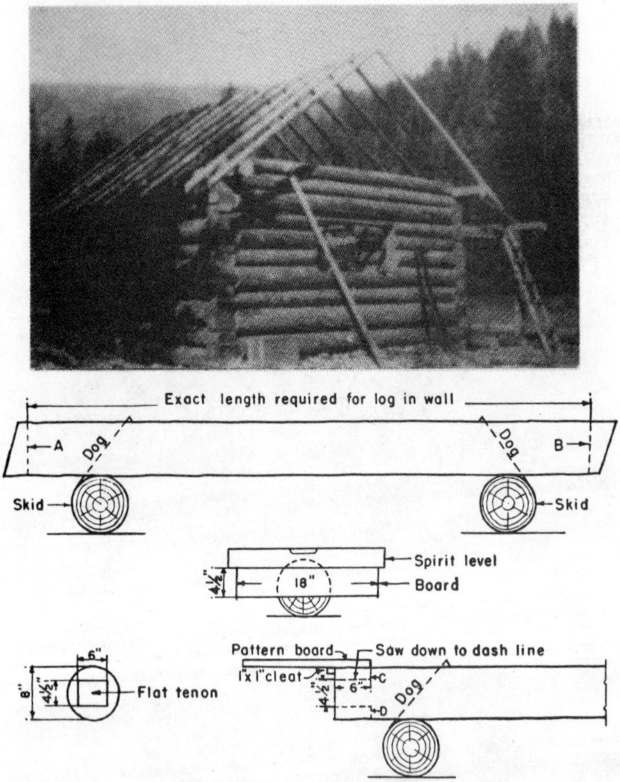

FIGURE 8.—Framing the flat, or plain, tenon corner.

edges. The width of a tenon varies with the diameter of the logs; 8- to 10-inch diameters will produce 6- to 7-inch wide tenons.

Nail a 1 inch by 1 inch cleat on the pattern board to points *C* and *D* and then make saw cuts on each end, cut chip off and smooth the surface. Turn log over and repeat on the other side. After framing out the sides of the tenon, the log is ready to be placed on the wall. Some fitting between corners is usually necessary but, if the logs are fairly straight and smooth, the work will be minimized.

The upright, or groove-and-tenon, corner (fig. 9) is used to a considerable extent in the West. It has desirable features from a mechanical standpoint: (1) The weight of the building is carried on the full length of the logs and does not rest solely on the corners, as in other types, and (2) it makes a tight wall because no openings

UPRIGHT OR GROOVE AND TENON CORNER

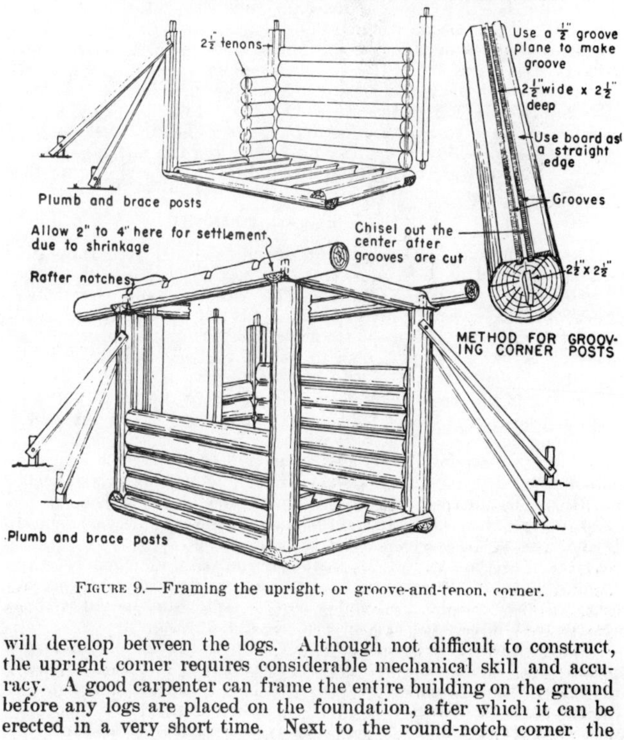

FIGURE 9.—Framing the upright, or groove-and-tenon, corner.

will develop between the logs. Although not difficult to construct, the upright corner requires considerable mechanical skill and accuracy. A good carpenter can frame the entire building on the ground before any logs are placed on the foundation, after which it can be erected in a very short time. Next to the round-notch corner the upright, or groove-and-tenon corner, probably has the best appearance.

DOOR AND WINDOW JAMBS

Door and window jambs should be framed just like the corners except that only the back should be grooved. The door side, or face, may be rabbeted or left smooth so that a separate wood door stop may be nailed in place. If the logs are reasonably dry, from 3 to 4 inches should be left at each corner for settlement due to shrinkage; otherwise, more or less space should be allowed, as conditions require. In about 6 months the cap log will come down and close this gap. Similar provisions should be made for settlement over door and window openings.

FLOOR JOISTS

As soon as the first round or tier of logs is laid, the floor joists should be set in place, notching them into the bottom side logs. If the building has a continuous masonry foundation, the joists may be set on top of it, as in a frame building.

In order that the ends of the joists may have sufficient bearing on the wall, it is necessary either to notch the ends into the side logs or hew the latter off on the inside. A simple method is to cut the notches in the side logs before they are rolled into place. Pole joists should be from 4 to 8 inches in diameter and hewed level on the upper side to provide a solid bearing for nailing the flooring. Several different ways of framing the floor joists are shown in figure 10.

LAYING THE WALL LOGS

In laying the successive rounds of logs in the walls, several details must be observed to keep them lined up so that the top logs form a level seat for the roof framing. The corners should be kept as level as possible as each round is laid. This can be done by measuring vertically from the top of the floor joists, from time to time, as a check. A variation of 1 inch in height will not cause a serious difficulty.

The height of the corners is regulated in two ways: (1) By increasing or decreasing the depth of the notch, and (2) by reversing the top and butt ends of the logs when laying them in the wall.

The logs should be fitted together as tightly as possible. In the case of somewhat irregularly surfaced logs, it may be necessary to smooth off certain portions of the under side of the upper log to secure a tight fit. Only in exceptional instances, however, should this be done to the top of the lower log.

The face of the logs on the inside of the building must be kept plumb, that is, in the same vertical plane. An ordinary carpenter's, or spirit, level may be used, but a 6- to 8-foot plumb board is considered most satisfactory because of its greater length.

The logs should be pinned together with a wooden pin or large spike (fig. 11). Spiking is done by boring a ¾-inch hole halfway through the upper log and continuing with a 7/16-inch hole through the bottom half. Then drive a 10- or 12-inch spike into place, or until it penetrates half the next log below. The spikes should be

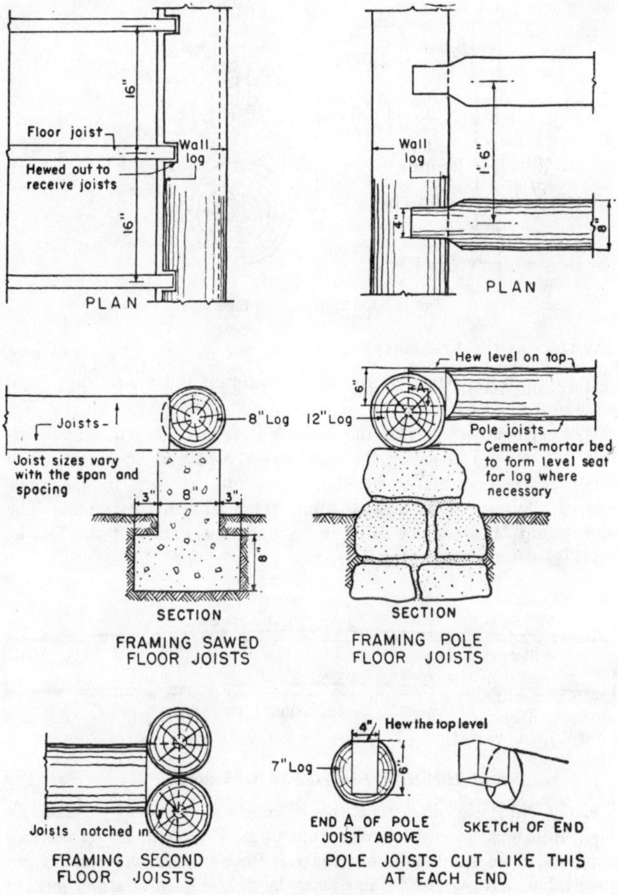

FIGURE 10.—Framing floor joists.

staggered in alternate rounds or tiers of logs. If wooden pins are used, fir or oak logs are preferable. Neither wooden pins nor spikes, however, offer interference to the settling of the walls.

The spike method is easier and quicker, and just as satisfactory as the wooden pin. The logs should be pinned approximately 2 feet from each corner and at each side of the window and door openings. For small structures, where the alinement of the walls is not so important, pinning may be eliminated, but it is essential to aline

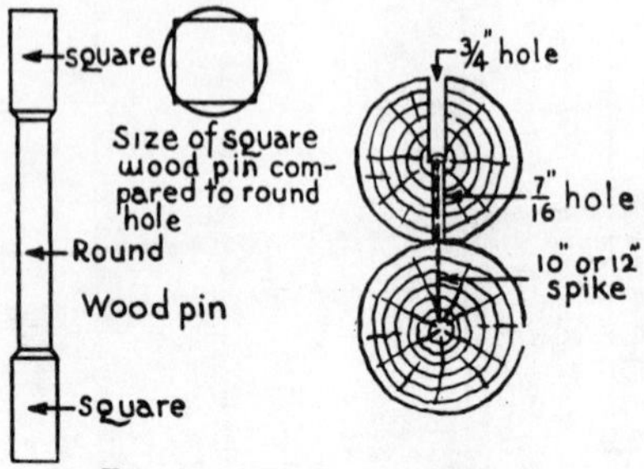

FIGURE 11.—Pinning logs together.

larger buildings accurately in order to prevent individual logs from springing out of place.

Where the use of logs having a decided curve, or sweep, is unavoidable they should be set in the wall with the bow or back up. Such logs may be straightened by making enough saw cuts in the upper side of the curved surface to allow them to straighten out. The cuts should be from one-third to one-half the depth of the log, or slightly more, if necessary (fig. 12).

FIGURE 12.—Straightening a curved log.

WINDOW AND DOOR OPENINGS

Early American log structures were characterized by relatively dark interiors because window openings, designed for protective purposes, were small and far apart. Since protection is no longer a consideration, window frames may be of standard size and located where they are most suitable for adequate day lighting.

As soon as the first round of logs and the floor joists are laid in place, mark the location of door and window openings on the inside face. Next saw out the door openings and chop out the notch in the doorsill log to within an inch of the true or finished line, as shown in figure 13. Leave final cutting of the openings to the exact dimensions until the window and door frames are to be placed in position, thus insuring a good finished wood surface. Also, determine the height of the openings above the floor line and mark them in figures

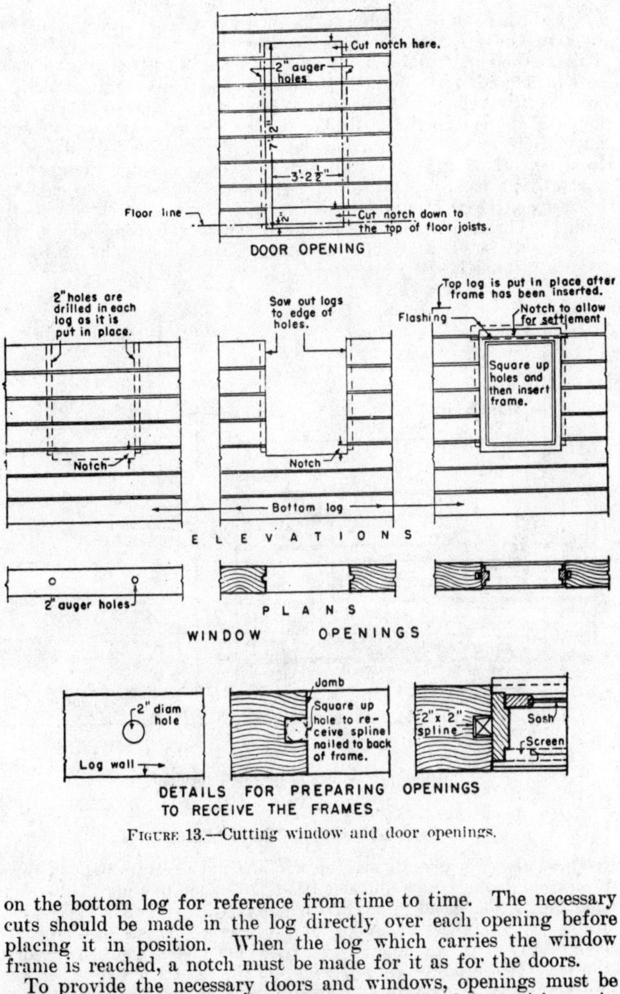

FIGURE 13.—Cutting window and door openings.

on the bottom log for reference from time to time. The necessary cuts should be made in the log directly over each opening before placing it in position. When the log which carries the window frame is reached, a notch must be made for it as for the doors.

To provide the necessary doors and windows, openings must be cut in the walls after the logs have been placed in position. As soon as a log in the wall is cut in two, the problem arises of how to hold the loose ends in place. Also, the doors and windows require the proper kind of frames to insure airtight closure between the latter and the ends of the wall logs. The most practicable and satisfactory method is to frame a vertical notch in the ends of the wall logs, into which can be fitted a spline attached to the back of the jamb or sidepieces of the door and window frames. This method of framing holds the wall logs in place, allows them to shrink and settle without hindrance, and makes a weathertight joint between them and the door and window frames. The vertical notch in the end of the wall logs may be framed by boring a 2-inch auger hole in each log as it is laid in place. The hole should be located so that, when the wall logs are sawed out for the opening, the saw cut passes down through the edge of the hole nearest the opening. It is then a simple matter to frame the notch to take the spline. The inside face of the notch can be left rounded and the spline chamfered to fit. To keep the holes in line from log to log, use the plumb board illustrated in figure 14.

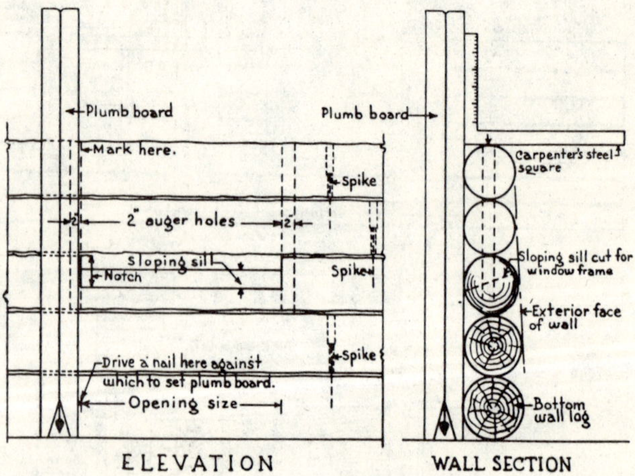

FIGURE 14.—Method of marking openings.

WINDOW AND DOOR FRAMES

There are two ways of making window and door frames—in three pieces (two side jambs and one head jamb), or in four pieces (two side jambs, one head jamb, and a sill piece). When a three-piece frame is used, the bottom log of the opening is cut or shaped to make the window or doorsill and the jamb pieces are then fitted to the sill. If the jambs are framed from pieces of log slabbed on two opposite sides, a presentable frame in keeping with the log character of the structure is obtained. The window or door face of the jamb pieces may be rabbeted for the windows and doors, respectively, or they may have separate wooden pieces, known as stops, nailed on. The spline on the back of the jamb may be rabbeted out, or a 2 inch by 2 inch piece of straight-grained wood nailed on. The head jamb can be framed in the same way; it does not require a

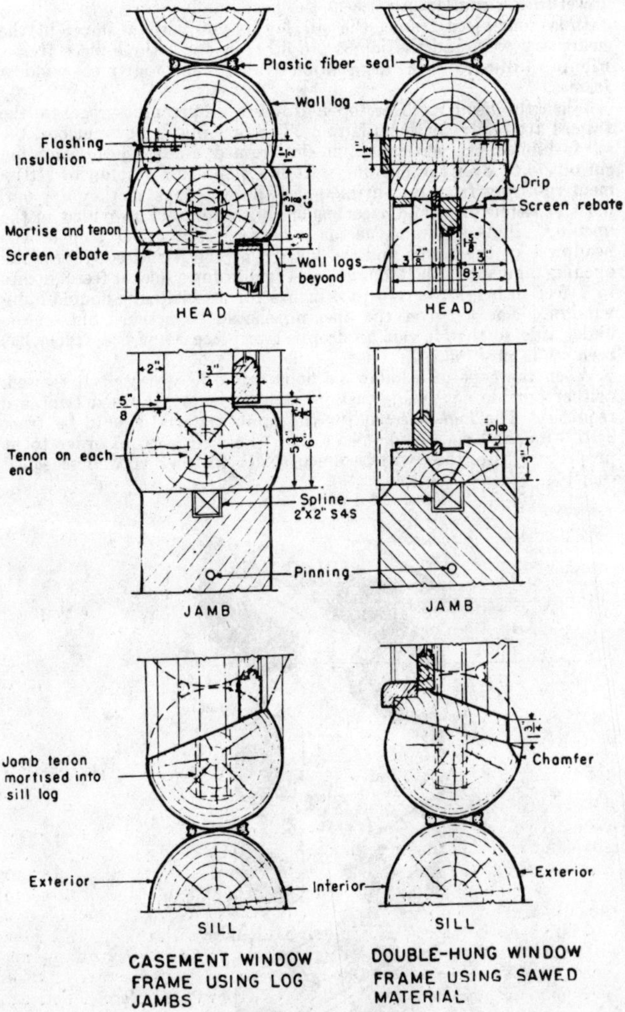

FIGURE 15.—Window frames.

spline on the back. Each side jamb has a dowel framed on each end. The bottom dowel fits into a mortise in the sill and the top dowel into a similar mortise in the head jamb.

In a four-piece frame, the sill log is cut with a slope, in the customary way, and the jambs are fitted as for a three-piece frame. Figure 15 illustrates the installation of three- and four-piece window frames.

When the head jamb or top log over the opening is reached, the frames are ready for installation. The opening is now cut out, the sill fashioned, the vertical spline slot framed, and the head jamb log cut out to fit over the opening. At this point, the amount of settlement resulting from the shrinkage of the wall logs, as they dry out, must be determined and a corresponding allowance provided in the opening. This allowance is made between the upper side of the headpiece of the frame and the bottom of the log directly over the opening, and should be from 2½ to 4 inches for a door 6 feet 8 inches to 7 feet in height, or 1½ to 3 inches for an ordinary double hung window. The log over the opening should be notched out on the under side so that it can be dropped in place after the frame has been set in position.

When the type of window or door frame here described is used, neither outside nor inside casings, sometimes called wood trim, are required. The logs selected for the jamb material should be from 2 to 3 inches larger in diameter than the wall logs, in order to fit properly. Also, they will be much easier to work if well seasoned (fig. 16).

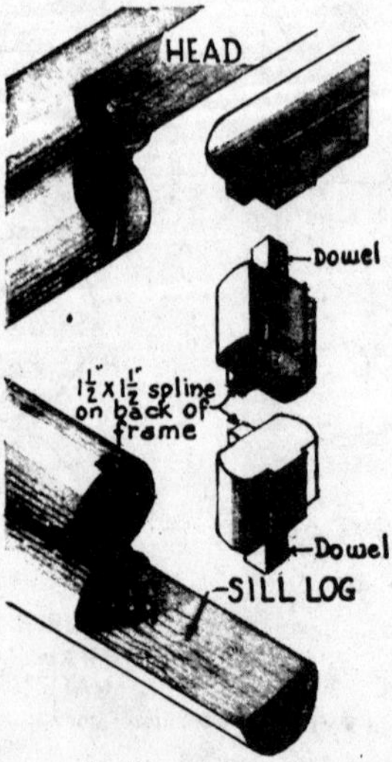

FIGURE 16.—Log jamb window frame.

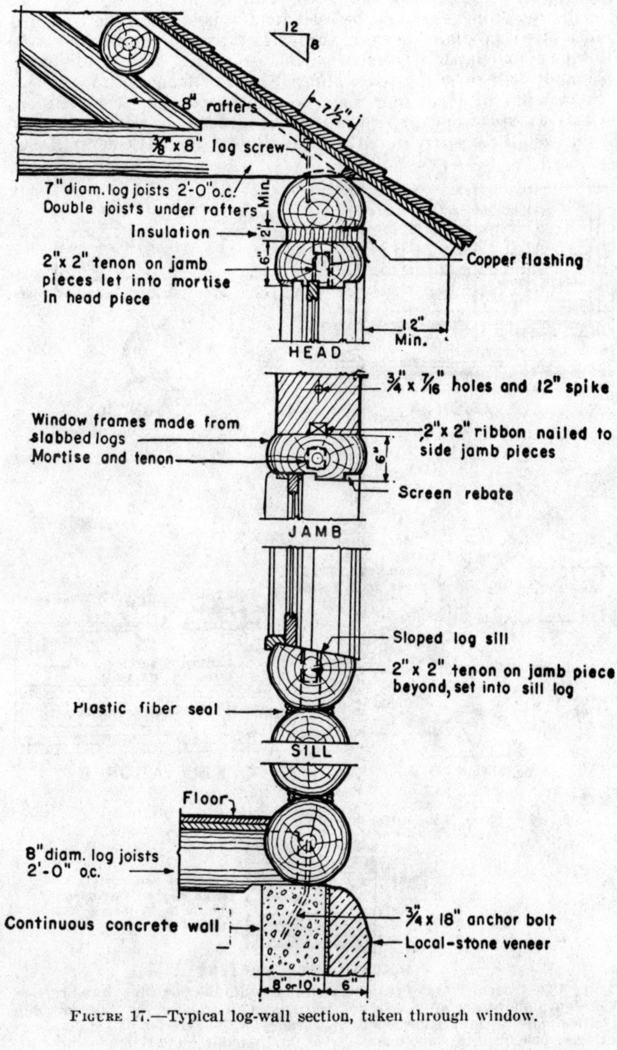

FIGURE 17.—Typical log-wall section, taken through window.

If standard millwork frames are used, false side jambs of sawed material, usually 2-inch planks, should be fitted in the openings to hold the logs in place. For a wall made of 10-inch logs, a plank 2 inches by 10 inches should be used for the jambs and the standard frame fitted in place between them after providing the necessary allowance for the wall logs to shrink or settle. The head casing ordinarily will cover the space allowed for shrinkage.

Some kind of insulating material which will take compression, such as crumpled newspapers, asbestos wool fiber, or rock wool, may be used to fill the space over the head allowed for settlement. Insu-

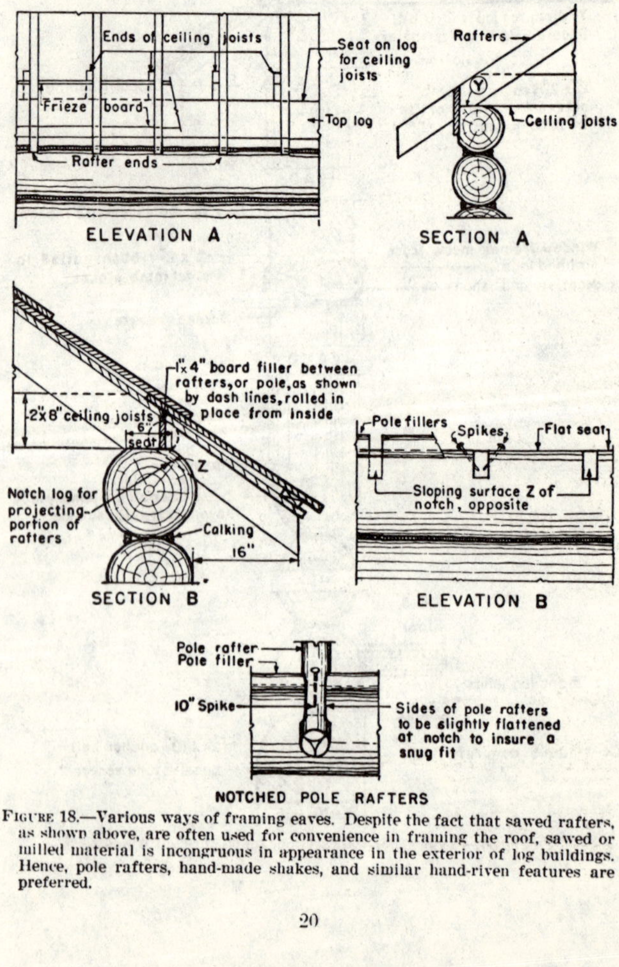

FIGURE 18.—Various ways of framing eaves. Despite the fact that sawed rafters, as shown above, are often used for convenience in framing the roof, sawed or milled material is incongruous in appearance in the exterior of log buildings. Hence, pole rafters, hand-made shakes, and similar hand-riven features are preferred.

lating material must be installed loosely, so as not to take any weight as the headlog gradually settles.

For the log-type frame, copper or galvanized steel flashing should be fastened to the bottom of the cut in the top log, leaving the lower edge of the flashing free to slide on the face of the log head jamb. As the wall settles, the bottom of the flashing can be trimmed off if too much of the face of the head jamb is covered. This makes a weathertight joint and protects the insulating material with which the shrinkage space has been filled. See figure 17, Head section.

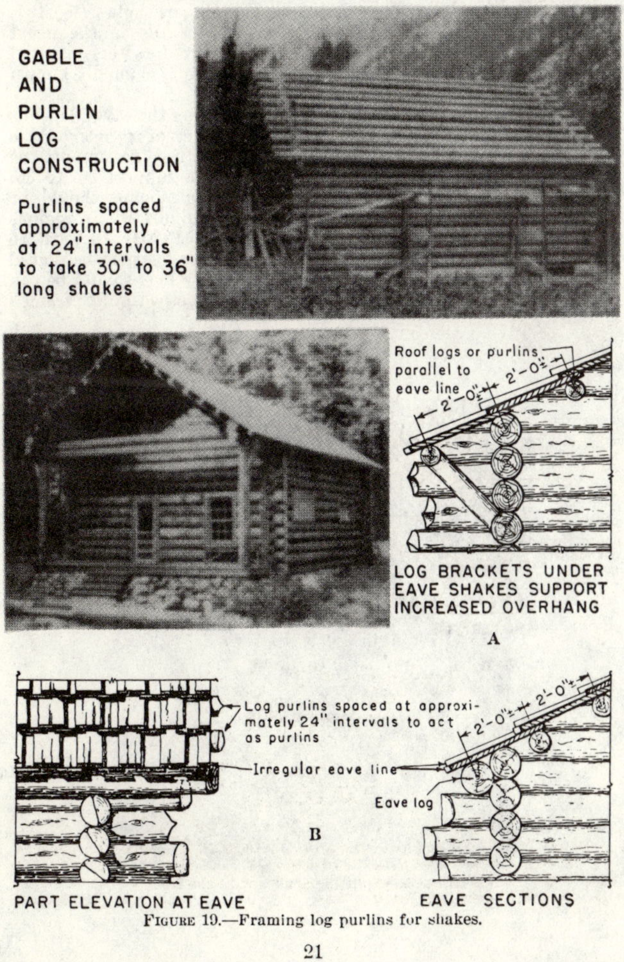

FIGURE 19.—Framing log purlins for shakes.

ROOF FRAMING

Roofs may be framed in several ways, depending upon the kind of material available and the appearance desired. The framing for a shingle roof, whether of sawed material or round poles, is done in the same way as that of a frame building. The top log on the wall may be cut with a flat seat for the rafters to rest upon, as at *Y*, in figure 18, *A* or notched out to receive them as at *Z* in figure 18, *B*. The gable ends may be run up with the logs, which is preferable for architectural appearance, or framed like the gables of a frame structure, and then covered with wood siding, shingles, or shakes (fig. 19).

The shingles may be laid over sheathing boards in the usual manner or on shingle strips placed across the roof rafters, parallel with the ridge and exactly spaced to receive them, commonly known as "barn-fashion."

The particular method to be followed in framing the eaves depends largely upon their projection. Where the effect of a considerable overhang is desired, an eave purlin log may be used to support the projecting shakes as shown in figure 19, *A*. To support 30- to 36-inch long shakes having a 6-inch lap, the log purlins should be spaced at approximately 24-inch intervals, as in figure 19. In regions of heavy snows, the eave log may be placed slightly forward to help support the overhang, or an additional eave log may be placed in position, as shown in figure 19, *B*. The gable logs should be run up at the same time as the roof logs, and both rigidly framed together.

FIGURE 20.—Splitting shakes with the froe.

Shake Roofs

It is often desirable to use hand-split shakes for the roof covering. These are usually made from cedar, but may be of any straight-grained wood, free from knots, which splits easily. First, the logs are cut in lengths of 30 to 36 inches and then the shakes are split off with a tool called a froe (fig. 20).

After the log cuts are set on end, the froe is held on the upper end of the block and then struck a blow with a wooden maul which causes a piece of the block or shake to split off. Being hand-split, the thickness varies somewhat; the minimum is ½ inch. A roof of thin shingles, lacking sufficient scale, is never as effective as a rough textured one, using ¾- to 1¼-inch thick shakes, to harmonize with the sturdy appearance of the log walls. The width, normally 6 to 8 inches, is governed by the size of the blocks of wood and varies accordingly, while the length is governed by the spacing of the roof logs or purlins. Shakes are always laid on the purlins in single courses, lapping the sides 1½ to 2 inches and over-lapping the ends at least 6 inches, as illustrated in figure 19. Nailing is usually done with six- or eight-penny galvanized box nails. Copper nails may be used for greater permanence. A good shake roof will not leak although from the inside of the building it may appear to have many holes.

The ordinary, uninteresting, straight-line effect at the butts may be broken up by staggering them from 1 to 2 inches, as is often done with shingles. This method produces an effect more in keeping with the log walls. Although involving greater care and additional labor it is preferable, from an architectural point of view, to the more common custom of laying them to uniformly straight lines.

At the ridge of the roof, where the shingles or shakes intersect, provisions must be made for weatherproofing. The shingled Boston ridge, comb intersection, or pole ridge, shown in figure 21 are practical and much more satisfactory from the standpoint of architectural effect than stock metal ridges, ridge boards, and other methods.

PARTITIONS

If the log building is to be divided into several rooms, at least two different methods may be used to construct the partition walls. If the log-construction plan is to be carried throughout the structure

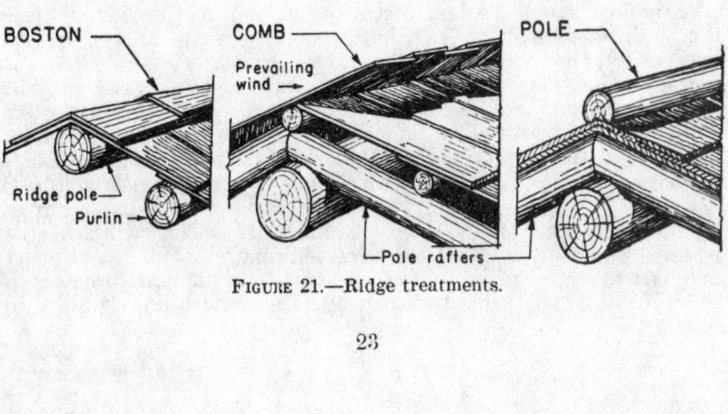

FIGURE 21.—Ridge treatments.

by using interior log-wall partitions, these should be laid out and framed in, and the door openings cut in the same manner as previously described for exterior walls. If a log partition comes at a place in a cross wall where it is not considered desirable to have the log ends project into the room beyond the opposite face of the wall, they may be sawed off flush with the face of the cross wall, as shown at *X*, figure 22, Plan *A*. This will not weaken the joint since the logs are both pinned and locked in place.

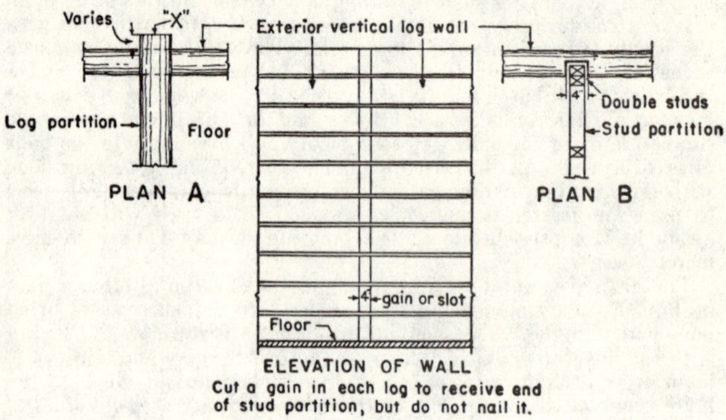

FIGURE 22.—Interior partitions.

Where frame partitions are used, they should be constructed as in a frame building. A gain or a 3- to 4-inch deep groove should be cut in the log wall into which the end studding of the frame partition is to be set (fig. 22, Plan *B*). The cut should be made in each log before it is placed in the wall. In no case should the studding at the ends of the partitions be nailed to the log walls which they intersect in order not to interfere with or be affected by their shrinkage and settlement.

FLOORING

A subfloor should be laid first using shiplap or sheathing. Over this a finished floor of such hardwoods as maple or oak, or the harder softwood species such as Douglas-fir, western larch, or southern pine, may be laid. Vertical grain and flat grain may be had in both softwood and hardwood, but the vertical grain shrinks and swells less than the flat, is more uniform in texture, wears more evenly, and the joints open much less. Finished flooring consists of tongue-and-groove material of various thicknesses and widths.

Despite a slight tendency to splinter and wear irregularly over a period of years, plain wide planking of random-width boards makes an appropriate floor for a log building. An attractive effect may be had by using screws instead of nails, countersunk to a depth of ½ inch and concealed by inserting false wooden dowels glued in place as shown in figure 23, *B*. Keying the boards together with wood keys, at random along the edges, adds to the attractiveness of the flooring.

INTERIOR WOOD FINISHING

Hanging doors and windows, and many other customary details of building construction should be done in the usual manner in building with logs. Whenever cupboards or other built-in units are constructed, they must be framed to be independent or entirely free of the log walls, like the furniture. However, such fixtures as lavatories may be attached to two adjacent logs without any subsequent structural complications.

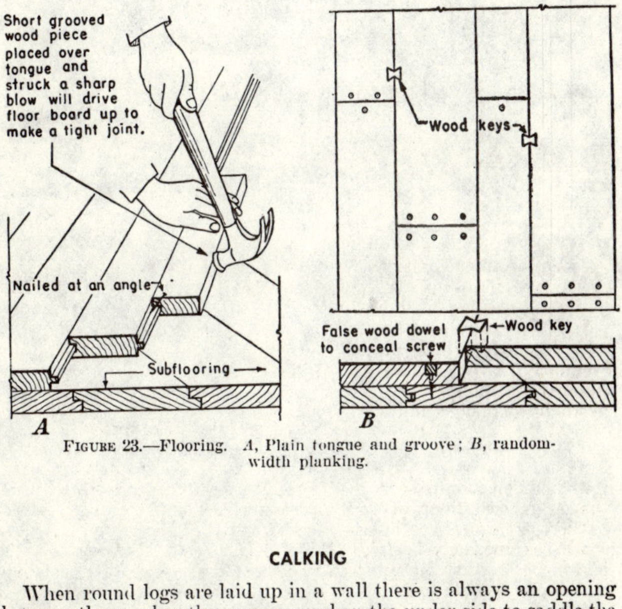

FIGURE 23.—Flooring. *A*, Plain tongue and groove; *B*, random-width planking.

CALKING

When round logs are laid up in a wall there is always an opening between them unless they are grooved on the under side to saddle the one below, as described later under chinkless log-cabin construction. In exterior walls, this opening, or crack, must be closed in order to make the structure weathertight. There are several methods of doing this. If the logs are reasonably straight and uniform in size and the corners carefully made, the opening between them will be small, often barely perceptible. When this is the case, the openings should be filled with some sort of calking compound applied with either a pressure gun or a trowel (fig. 24).

In recent years several kinds of calking material have been put on the market. They are applied best with a gun having a pressure-

FIGURE 24.—Examples of tight joints well calked. *A*, Interior calking; *B*, exterior calking.

release trigger whereby the calking compound is forced through a nozzle made in various shapes and sizes to meet different requirements. These calking compounds are not adversely affected by heat or cold, retain their natural flexibility, and have an adhesive property which causes them to adhere to the surface to which they are applied.

A good plastic compound will adhere to the logs under all conditions and can be patched easily by simply applying more material. A black fiber seal is not objectionable and, at the same time, gives a practical finish. The seal should be applied to both sides of the exterior and interior log walls, producing an almost hermetically sealed building. When applied with a pressure gun having a 3/8-inch nozzle, 1 gallon will fill about 300 linear feet of opening. If applied in cold weather, the material should be heated to a temperature of 60° F.

CHINKING

When using logs that are somewhat rough and irregular in shape, the resulting space between them may be so large that the calking material cannot be used satisfactorily to fill the opening. In such cases, it will be necessary to insert "chinking," which usually is applied to the interior and exterior walls in one of two ways:

1. *Split chinking.*—Segments of a log are split out in sizes which fit the opening and, after being carefully shaped with the ax to make a tight fit, are securely nailed in position. This kind of chinking requires considerable work and patience to secure a good appearance.

2. *Pole chinking.*—Small round poles may be used to fill the openings (fig. 25). Usually they are cut in sizes and lengths to fill

FIGURE 25.—Pole chinking.

the opening from wall to wall. This sort of chinking may be applied rapidly to either inside or outside walls and makes a neater job than the preceding method. Unless the logs are thoroughly seasoned these small poles sometimes have a tendency to pull away from the nails. When the chinking has been completed, the openings will have been reduced sufficiently in width to allow the calking material to be applied successfully.

It is always a serious problem in log construction to devise a practical method for permanently fastening the plaster daubing in place on both inside and outside walls. In some instances, shingle nails may be driven into the logs 2 to 3 inches apart for the full length of the opening or 2-inch wide strips of metal lath may be used and the plaster applied to fill it. Cattle hair may be added to the plaster to increase its adhesive consistency and thereby hold it more rigidly in place. Sometimes, wood strips are nailed on the lower log to hold the plaster in position, as shown in figure 26, but they are unsightly.

CHINKLESS LOG-CABIN CONSTRUCTION

Chinkless construction, associated with the building of log structures in Scandinavian countries, eliminates the chinking and mudding so prevalent in many log buildings. It consists of grooving

FIGURE 26.—Wood daubing strips.

the under side of every log in each tier so that it saddles the log beneath, making a close joint for its entire length. The groove is marked by a tool which, for convenience, may be called a cabin scribe or a drag (fig. 27).

Directions for chinkless log-cabin construction.—Mark and cut out the notch just as is done for a round-notch corner. Next, dog the log in place and scribe, making the additional mark shown by dash line (X, fig. 27). Then, cut to line and, finally, drop log in position.

The scribe is 12 inches long, made preferably of 3/8-inch square steel or iron bent in much the same manner as the spring in a steel trap; the two ends are turned down about 1½ inches like two fingers, diverging to about ¾ of an inch at the points, and then sharpened with a flat surface on the inside of the point toward the loop. The loop should be hammered out thin to provide sufficient flexibility to allow the points to spread or close easily. A ring is welded around the two halves of the tool which, when slipped up or down, makes it possible to adjust the points and thereby prevent any further spreading while the tool is in use. A link from a small chain, placed over the legs before the points are turned, will serve the same purpose and, to prevent the points from springing together, a small piece of wood may be forced between them.

To fit a log, first frame it at the ends and then fit it down to within about 2 inches of the lower log where the opening is the widest. It is difficult to do a good job of scribing when the logs are too close together. The scribe must then be adjusted at the point where

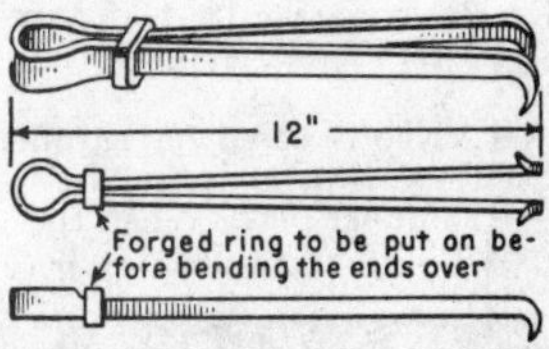

NORWEGIAN LOG SCRIBE

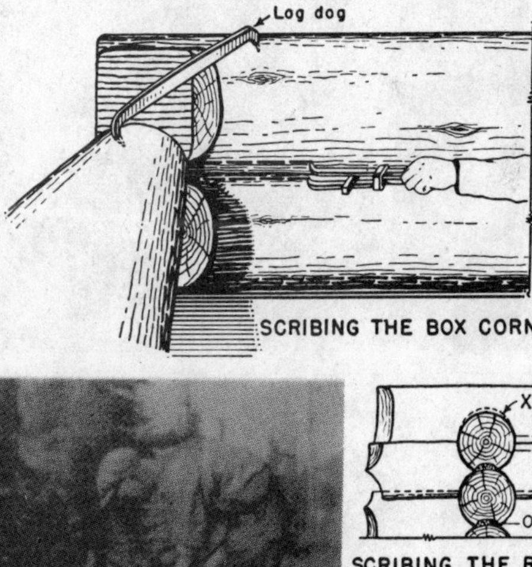

SCRIBING THE BOX CORNER

CHOPPING THE GROOVE

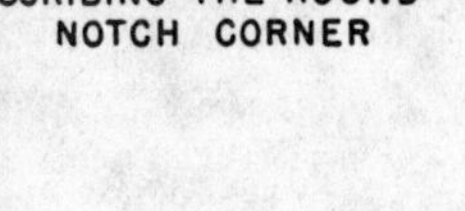

SCRIBING THE ROUND NOTCH CORNER

FIGURE 27.—Chinkless log-cabin construction.

the opening is the widest so that, when holding the tool parallel to the opening, the lower point of the scribe will ride on the surface of the bottom log. By exerting sufficient pressure, the upper point will score the top log. Repeat this operation to score the upper log on the other side. The corner tenons must be marked likewise. Next, turn the log over, work the tenons down and then cut a **V**-shaped groove to the marked lines in the remaining portion of the log, using a double-bitted ax. This groove should be cut deep enough along its center to permit the outer edge of the groove to rest continuously on the lower log. By removing the least amount of wood to make the smallest possible groove, the closest fit is obtained with the least effort.

The principle of the scribe is based on parallel lines, and it can readily be seen that if there is a hump on the lower log there will have to be a gouge in the upper one. When the work is done carefully, the space remaining is negligible. Where an airtight wall

FIGURE 28.—Fine example of milled-log construction—ranger's dwelling, Whitman National Forest, Oreg.

is desired, a strip of plumber's oakum should be laid on the bottom log before the upper log is dropped into place. If this material is not available, dry moss is a fairly practical substitute.

Milled-Log Construction

Sometimes it is feasible to take advantage of a portable mill to face the logs on three sides rather than to hew them by hand. The level beds seat the logs so well that calking is minimized, the smooth interior surfaces permit of easy finishing, particularly where wood wainscoting or plaster is used, while the round-log exterior effect is undisturbed, except where the logs project at the corners. Figure 28 illustrates a structure built in this way.

HEWING TIMBERS

The facing or hewing of round timbers to obtain one or two sides surfaced flat for framing purposes, as shown in figure 29, requires considerable skill in the use of the ax and broadax. There are, however, a number of mechanical aids (fig. 30) which should be used by anyone undertaking log construction in order to simplify the work as much as possible. The carpenter's spirit level, the steel square, and chalk line and chalk are necessary for laying off the lines to be followed in hewing timbers. In framing logs they should be laid up on skids, or sawhorses, dogged fast in place with iron dogs, and the dimensions laid off on each end of the log with the level and square to insure that the lines are parallel to each other. Then, with the chalk line, carefully snap lines on the side of the log connecting corresponding points at each end. For squaring the ends of a log and cutting pole rafters, use the miter box to guide the saw. To measure lengths accurately the steel tape, or a board pattern cut to the exact length, may be used.

FIREPLACE FRAMING

The living-room fireplace, invariably the most prominent interior feature, harmonizes best with a log interior if built of stone and provided with a crude log shelf. The fireplace itself may be either the traditional masonry type or the more modern metal-lined one equipped with a heatilator.

The masonry of the fireplace and its chimney should always start on solid earth, below the frost line, like the foundations of the building itself. Masonry does not settle, unlike the surrounding log construction. Consequently, it is recommended that a self-supporting log framing be built around and entirely free of the masonry of the fireplace and chimney, as illustrated in figure 31. The opening should be framed in the same way as window and door openings. The fireplace and chimney masonry should not be erected until the opening has been framed for it. Upon completion, the intersection between the stone and wood should be thoroughly calked to make an airtight, weatherproof job. This method allows the wall logs to settle, because of the unavoidable shrinkage, without structural failure.

FIGURE 29.—Framing hewed timbers.

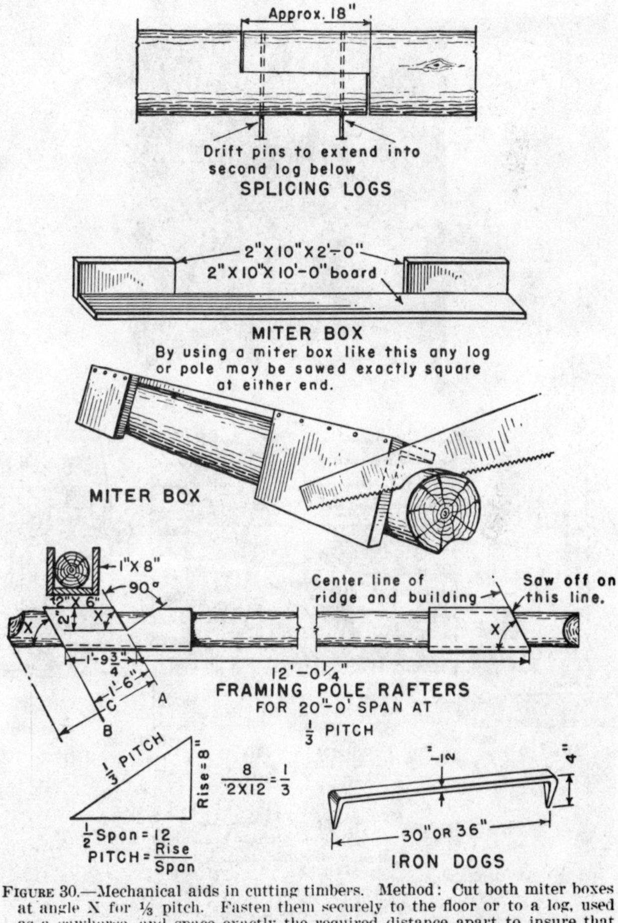

FIGURE 30.—Mechanical aids in cutting timbers. Method: Cut both miter boxes at angle X for ⅓ pitch. Fasten them securely to the floor or to a log, used as a sawhorse, and space exactly the required distance apart to insure that all rafters are cut alike. Then place each rafter in the boxes, back down if any curvature exists, dog rigidly in place and saw to the pattern.

Line A represents the exterior wall face and, if sawed off on line B, parallel with the wall face, overhang of eave will be 1 foot, 6 inches. Any desired overhang may be had and sawing eliminated by fixing the distance C. The irregularly hewed rafter end is preferable to the uniform elliptical saw-cut ends. Finally, hew the upper surface of the rafters to a smooth even bearing to receive the roof sheathing boards.

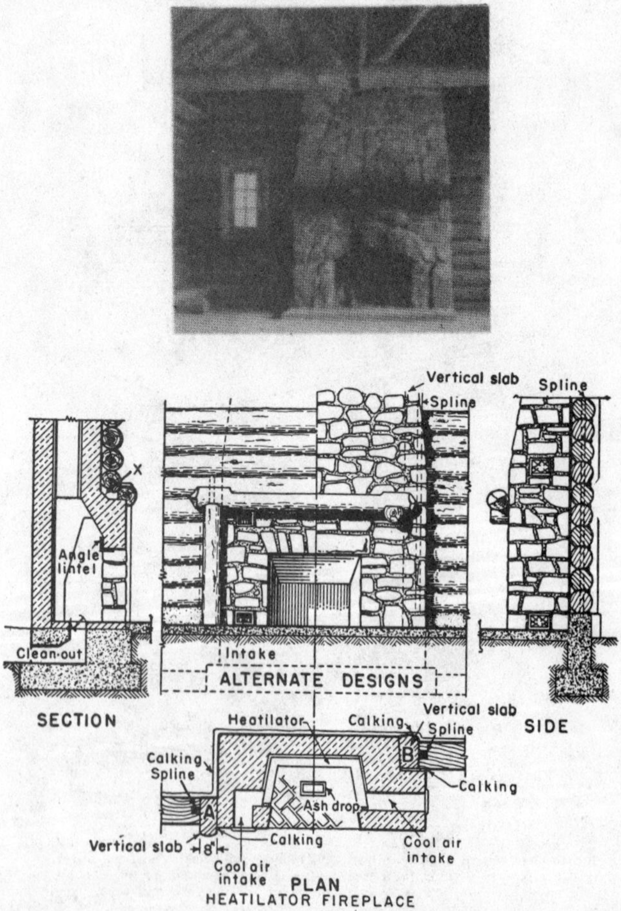

FIGURE 31.—Framing around the fireplace. Framing logs around fireplace and chimney varies with the effect desired: (1) By using an exposed vertical slabbed log and spline, as at A, with space X, to allow for the shrinkage settling of the logs above the mantel, or (2) by using a concealed vertical slabbed log and spline, as at B, where the masonry is exposed above the mantel.

FIGURE 32.—A useful type of modern log dwelling—ranger station, Gallatin National Forest, Mont.

In building an ordinary fireplace, the firebox and inner hearth should be made of firebrick to withstand intense heat and the various parts proportioned in accordance with standard practice to insure efficient operation.[1]

The heatilator is a built-in recirculating steel unit consisting of metal sides and back to form a heating chamber, adjacent to the fire pit, which draws cold air through a register at each side near the floor and after the air is heated ejects it through similar registers above. It should be installed in conformity with the manufacturer's directions, taking care to select a stock-size unit suitable for the dimensions of the fireplace opening and to erect the surrounding masonry accordingly.

OILING AND PAINTING

After all the openings have been properly calked and the logs brushed clean, it is often desirable, although not absolutely necessary, to treat the log surfaces with some sort of preservative material. Logwood oil is excellent for the exterior. The colorless variety is preferable in most cases but, if some color is desired, add just enough burnt umber, or raw sienna paste, to give the proper shade. For interior finish, apply a coat of clear shellac and then one or two coats of dull varnish. The trim can be treated in a similar manner to preserve the pleasing effect produced by the natural surface and color of the wood.

THE FINISHED STRUCTURE

Examples of modern log construction are shown in figures 32, 33, and 34. Early types of log structures are illustrated in figure 35.

[1] For this purpose the following publication will be found useful: Farmer's Bulletin 1889, Fireplaces and Chimneys.

FIGURE 33.—Modern structures showing effective use of log construction in recreation buildings on national forests in Montana. *A*, Dude ranch; *B* and *C*, recreation and mess hall, Seely Lake.

FIGURE 34.—Organization camp at Seely Lake showing log work in greater detail. *A*, Entrance wing; *B*, cabin group. Note the wedges under porch post to provide for settling of walls. Wedges are gradually driven out as necessary.

FIGURE 35.—Early types of log structures built by the U. S. Forest Service in the West. *A*, Ranger station, Gallatin National Forest, Mont.; *B*, ranger's dwelling, Nezperce National Forest, Idaho; *C*, log cabin in Arizona.

FURNITURE

The matter of interior furnishings is always of great concern to those who build log cabins. Odds and ends or too many "whatnots" may prove to be misfits. Pieces of Early American design are perhaps the most appropriate ready-made furniture, but sturdy, rustic pieces yield the greatest satisfaction.

Many cabin owners have found a great deal of pleasure in making essential furniture, such as bunks, beds, tables, chairs, settees, and similar items. In the East, birch is preferred as a material, and in the West, lodgepole pine is most satisfactory. Other native species, however, will do just as well. In making furniture it is advisable to remove the bark from the logs because bark collects insects, causes the wood to deteriorate and eventually falls off, leaving imperfect, unsightly surfaces. Figures 36 and 37 show types of furniture suitable for log residences.

For rustic effects, the use of a stain of the following proportions gives a satisfactory appearance: 2 quarts turpentine, 2 quarts raw linseed oil, and 1 pint liquid drier, to which add ½ pint of raw sienna, ½ pint of burnt umber, and a touch of burnt sienna. The top surfaces of tables, buffets, chests, and rawhide seats should have two coats of spar varnish. Where countersunk screws are used in connection with a stain finish, insert false wood, dowel-like plugs in preference to plastic wood to conceal the screwheads.

Simplicity, both in construction and appearance, is the keynote for producing the most harmonious effects in furniture, in keeping with log interiors.

Chairs and Stools

Armchairs can be built with well-seasoned lodgepole or eastern pine, or birch (fig. 38). The cornerpieces should be mortised and tenoned to the frame and rail and anchored in place with ⅜- by 6-inch lag screws. The arms should be fastened to the cornerpieces with ⅜- by 5-inch carriage bolts and to the slab support with ⅜- by 4-inch lag screws. The vertical slab support should be rigidly secured to the frame with ⅜- by 3-inch carriage bolts. Cushions may be of the filler type, without springs, and covered with homespun fabric. Use 2-inch wide heavy canvas strips, securely fastened with furniture tacks, to support the cushions.

Upright chairs and stools (fig. 39) can be made from the same material as the armchair. Cross the poles to impale the legs rigidly. The crosspieces of the chair back should be curved to fit the human back. The joints must be tightly glued, mortised, and tenoned.

Bed and Bunk

Birch or well-seasoned lodgepole or eastern pine is suitable for making a bed or bunk. In making a bed (fig. 40) the crosspieces

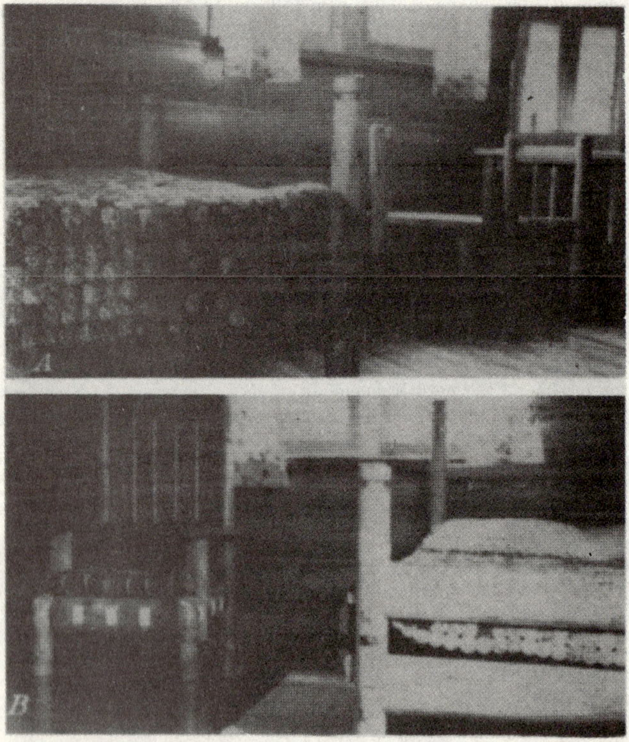

FIGURE 36.—Furniture suitable for log cabins—convenient, sturdy, and easy to make. *A*, Bed; *B*, bed and armchair.

FIGURE 37.—*A*, Dining table appropriate for log cabin; *B*, book rack and hod.

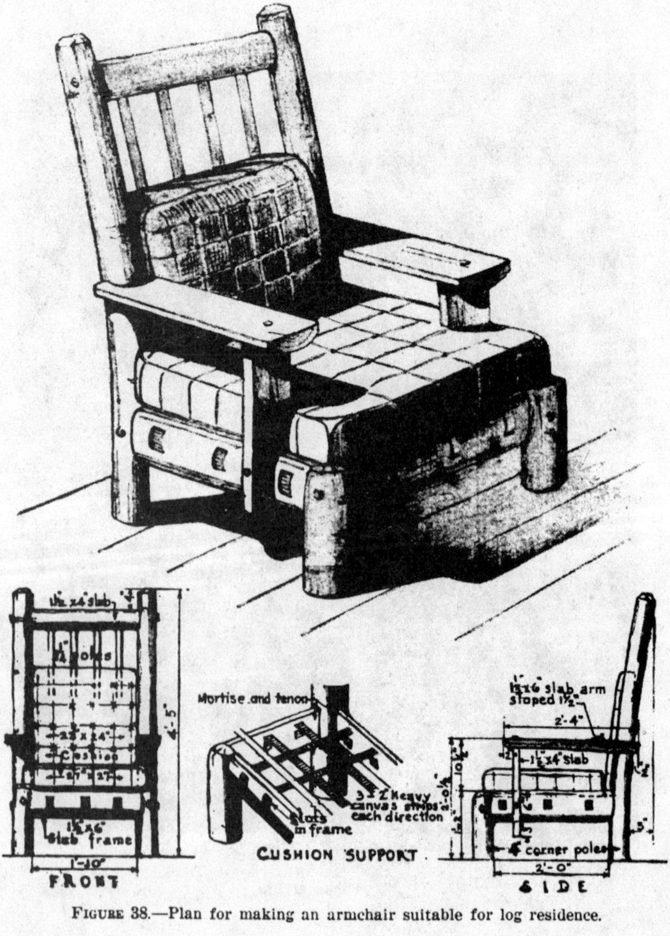

FIGURE 38.—Plan for making an armchair suitable for log residence.

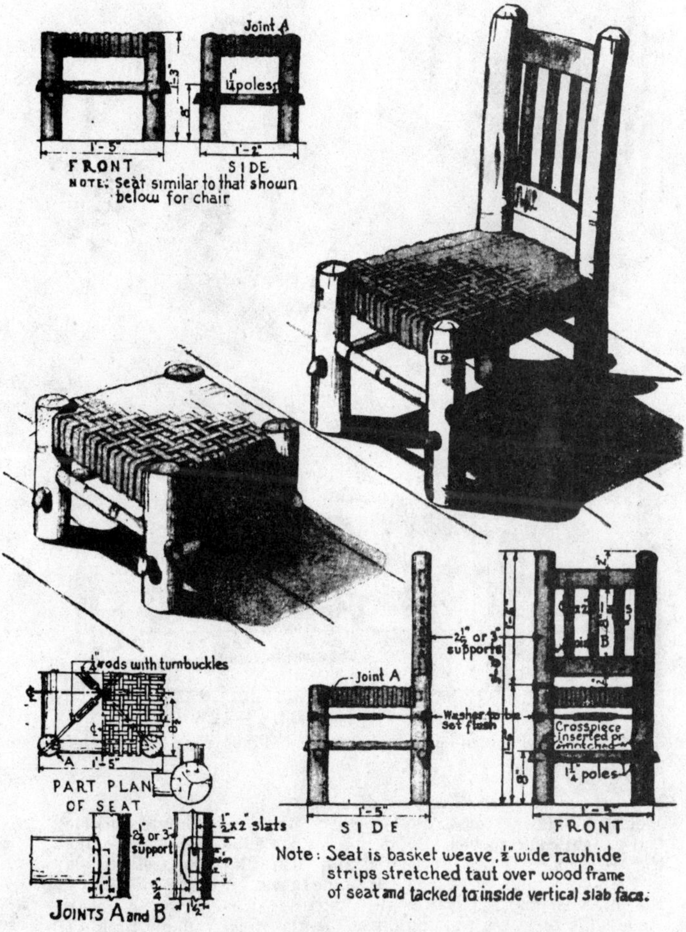

FIGURE 39.—Plan for making an upright chair and stool.

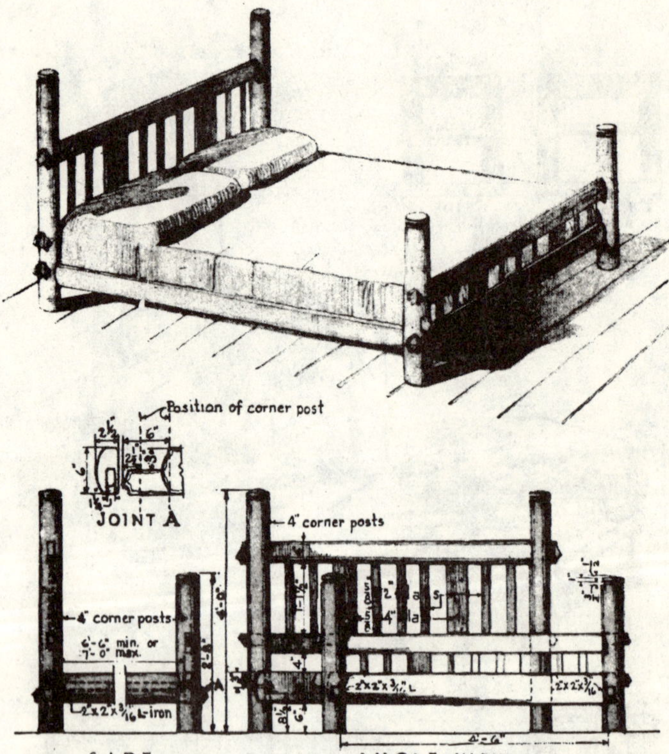

FIGURE 40.—Plan for making a double bed for log residence.

should impale the corner posts tightly; the joints should be glued and toe-nailed from below. Do not cut the side or end pieces until the bedspring has been measured and then allow for a slight play in both directions in setting the angle irons, in order to facilitate the insertion and removal of the mattress. Use 1/4- by 3-inch carriage bolts to fasten the angle irons to the wood frame. Figure 40 is a plan for making a double bed; for a single bed, reduce the width accordingly.

A double-deck bunk is made in much the same way as a bed (fig. 41).

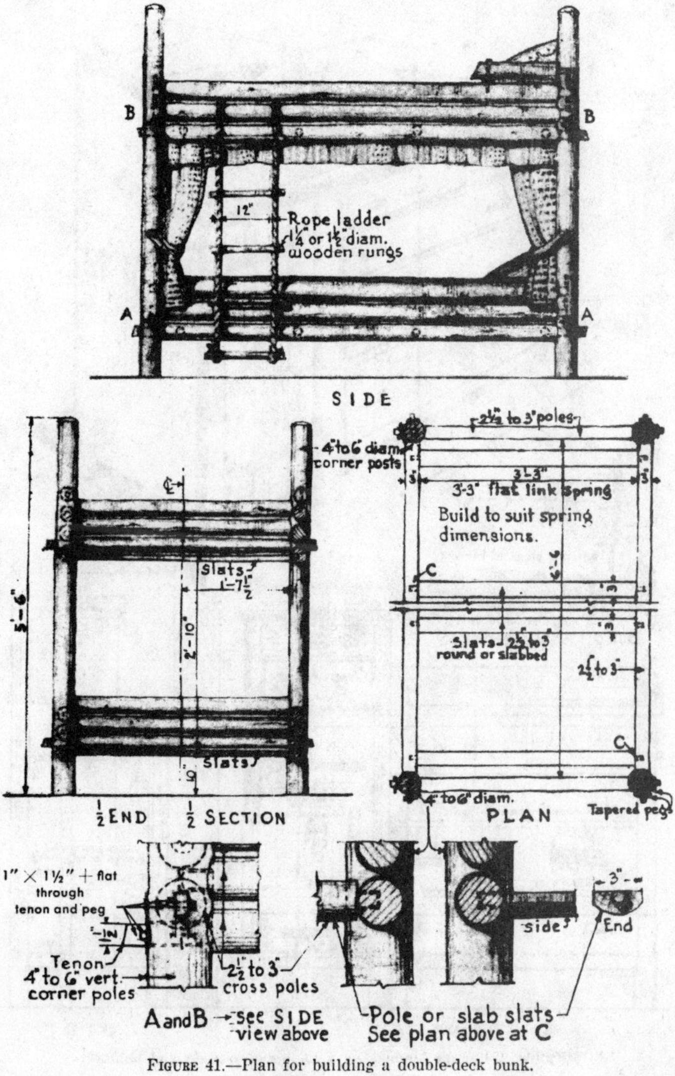

FIGURE 41.—Plan for building a double-deck bunk.

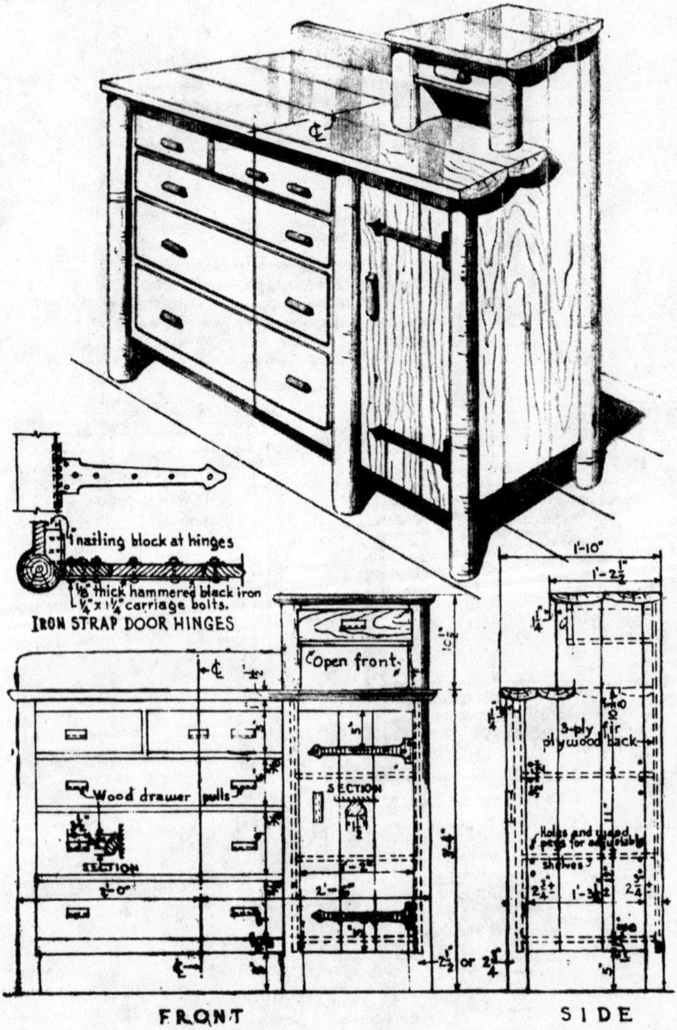

FIGURE 42.—Plan for making a combination chest and buffet.

Chest and Buffet

No log residence is complete without furniture for storing clothes. A combination chest and buffet suitable for log cabins can be made from well-seasoned lodgepole or eastern pine, tamarack, or birch (fig. 42). The ends, doors, shelves, and drawer fronts should be cut from No. 2 tongue-and-groove commercial pine lumber.

Settee

A settee can be made from well-seasoned pine or birch (fig. 43). Join the corner poles to the slab frame and rail with mortise-and-tenon joints; then anchor the joints by means of 3/8- by 6-inch lag screws. Fasten the arms to the corner poles with 3/8- by 5-inch carriage bolts and to the slab support with 3/8- by 4-inch lag screws.

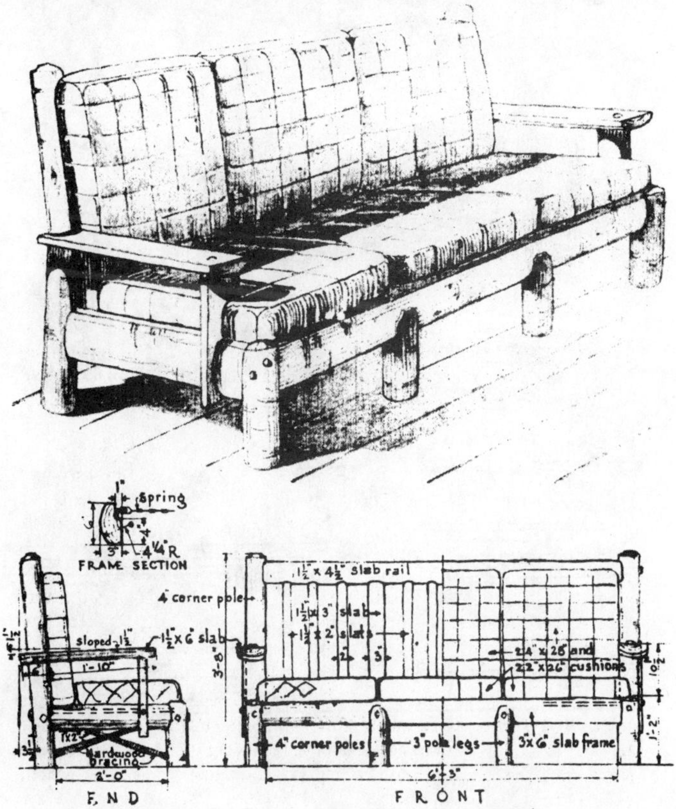

FIGURE 43.—Plan for making a living-room settee.

FIGURE 44.—Dining-table plan.

Use $\frac{3}{8}$- by 3-inch carriage bolts to fasten the slab support to the frame. The 1- by 2-inch hardwood crosspieces should be securely fastened at the top ends and notched into the legs at the bottom ends, held by 2-inch wood screws, driven into place at an angle. Back slats should be mortised and tenoned to the rail and frame. The cushions should be the filler type, without springs if so desired, and covered with homespun fabric.

Dining Table

Peeled pine or birch is ideal material for building a dining table (fig. 44). Make a tight saddle joint between *B* and the legs. Cross poles to impale the legs tightly. Notch *E* for the cross poles. Upper

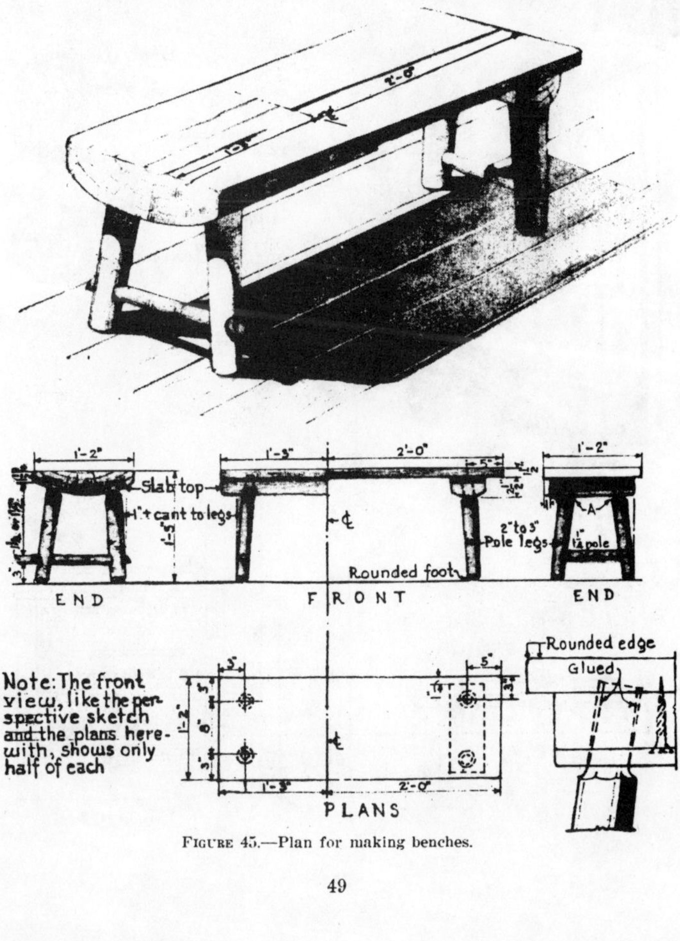

FIGURE 45.—Plan for making benches.

surface of *C* should be slab-faced and fitted between *D* and cross poles, all rigidly braced together. Top pieces of tables should be doweled at places indicated in the drawing with ½- by 4-inch wood dowels, glued and clamped to insure tight joints. Notch top pieces *A* 1-inch deep to receive *B* and *D*. Top outside edges of *A*, *C*, and *E* should be hewed.

Table, Bench, Book Rack, and Wood Hod

Well-seasoned lodgepole or eastern pine, tamarack, cedar, or birch are suitable for benches (fig. 45). The joints should be glued.

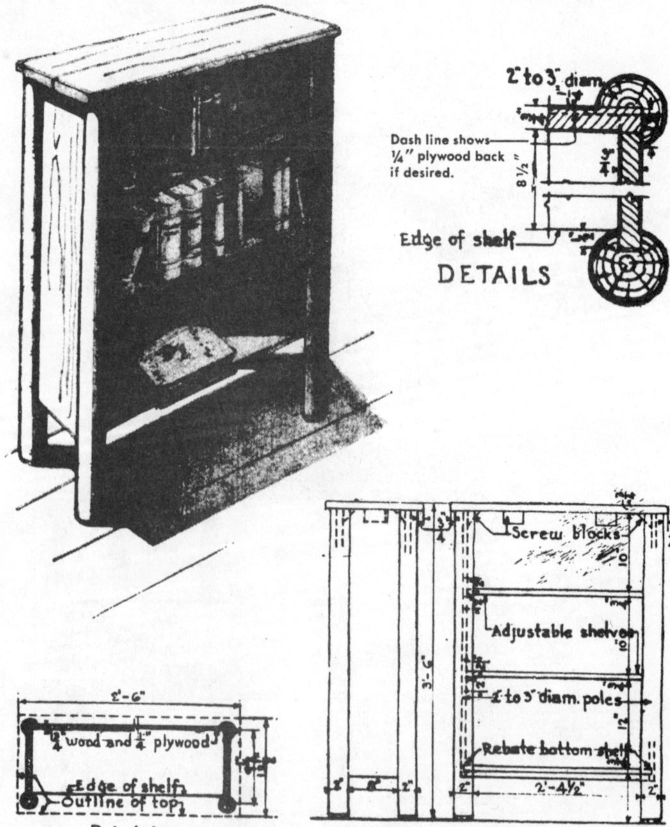

FIGURE 46.—Plan for a book rack.

Countersink any screws, then conceal the heads with false wooden dowel-like plugs. If the furniture is to be painted, use plastic wood. A book rack may be made of the same material used for the bench, except cedar, which is unsuitable (fig. 46). The sides and bottom shelf should be rabbeted and thoroughly glued. The two intermediate shelves can be made adjustable by boring 3 holes in each sidepiece 2 inches apart, above and below the position shown for

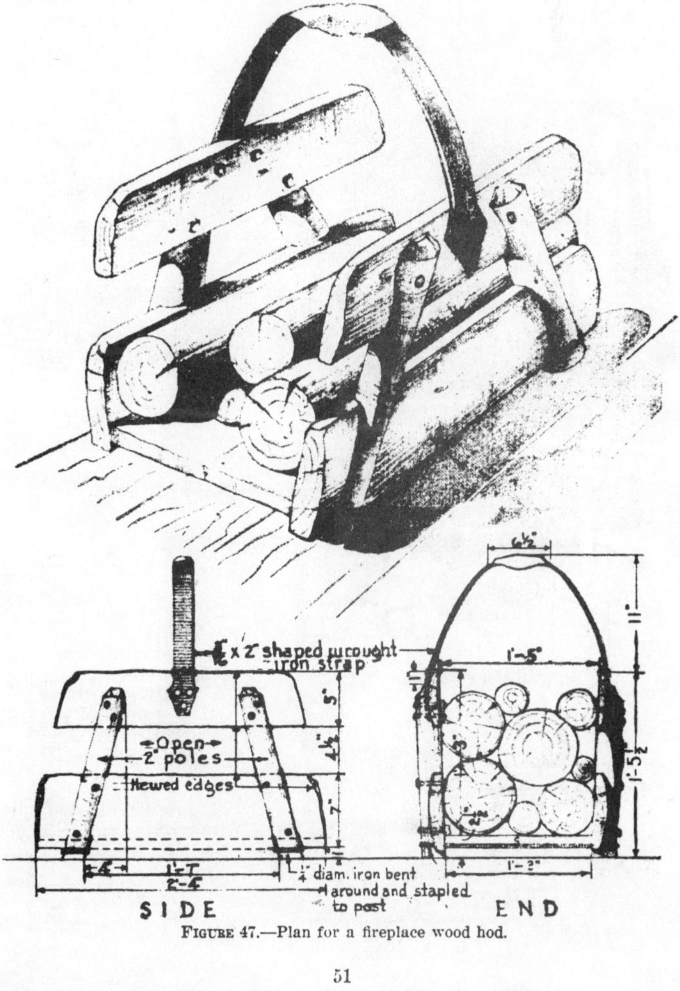

FIGURE 47.—Plan for a fireplace wood hod.

the shelves in figure 46, into which loose wooden pins may be inserted for their support. Screw the top in place, countersink screwheads and insert wood cover plugs or false dowels for concealment where stained finish is used. If painted, plastic wood may be used.

A fireplace wood hod (fig. 47) may be made of wood and metal. Use well seasoned lodgepole or eastern pine, tamarack, or birch. Make a tight cradle joint between horizontal and vertical side pieces, using 1/4- by 2-inch carriage bolts except that 1/4- by 3-inch lag screws

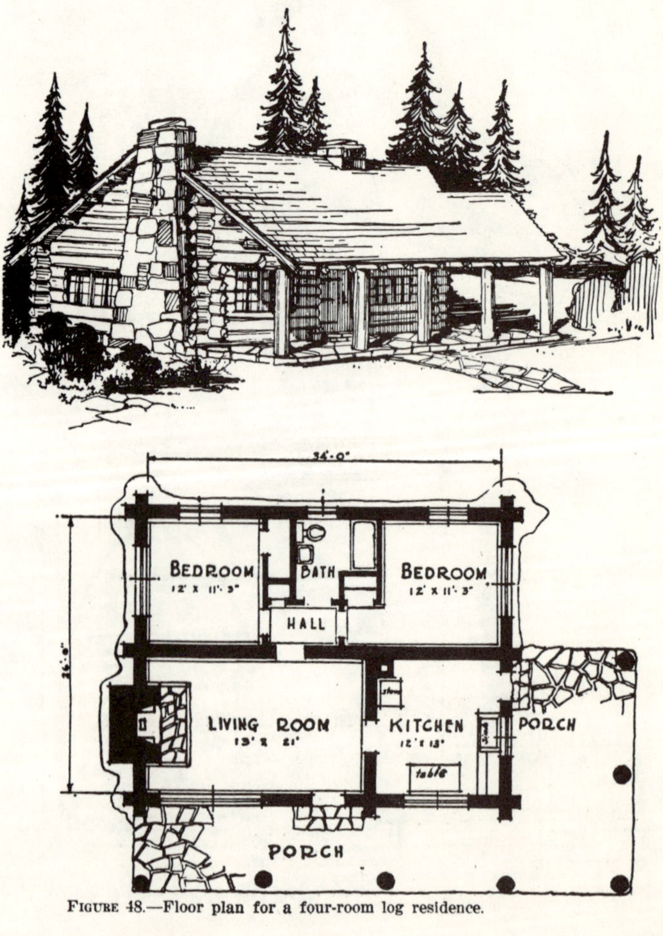

FIGURE 48.—Floor plan for a four-room log residence.

should be used for fastening the lower sidepieces and bottom. Secure the wrought-iron handle to each side toppiece with 3- by $1\frac{1}{2}$-inch carriage bolts. The wood sides should have hewed edges of $\frac{3}{4}$ inch minimum thickness.

BUILDING PLANS

Selection of the site and preparation of building plans varies with individual taste. In choosing a location one must consider avail-

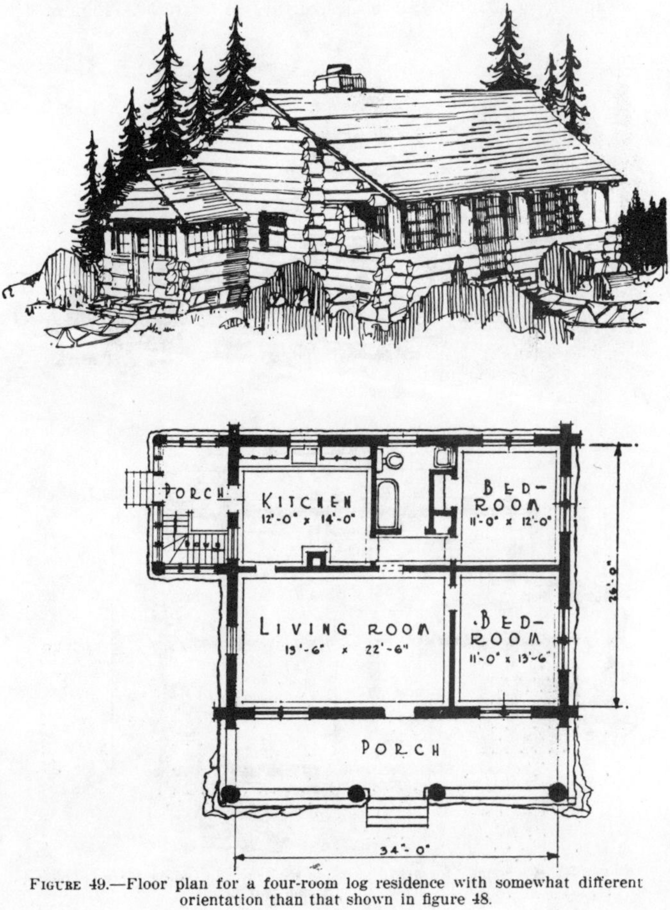

FIGURE 49.—Floor plan for a four-room log residence with somewhat different orientation than that shown in figure 48.

ability of transportation, shopping centers, water supply, sewage disposal, electric facilities, and kindred factors.

Before undertaking construction it may be desirable to consult an architect or competent builder to make sure that (1) your desires are satisfied with respect to the necessary accommodations; (2) rules and regulations enforced by local authorities will be observed; and (3) provisions are made for installing telephone, electricity, water, and plumbing facilities. Failure to take these precautions may necessitate costly changes after construction has begun.

Plans for suitable four-room log residences are given in figures 48 and 49, and for a five-room structure in figure 50. Figure 51

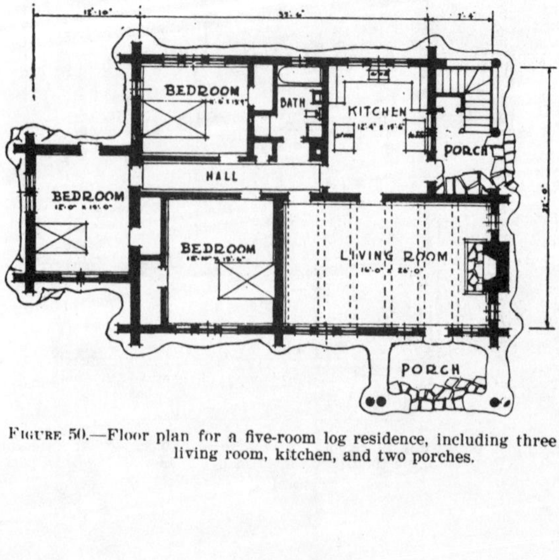

FIGURE 50.—Floor plan for a five-room log residence, including three bedrooms, living room, kitchen, and two porches.

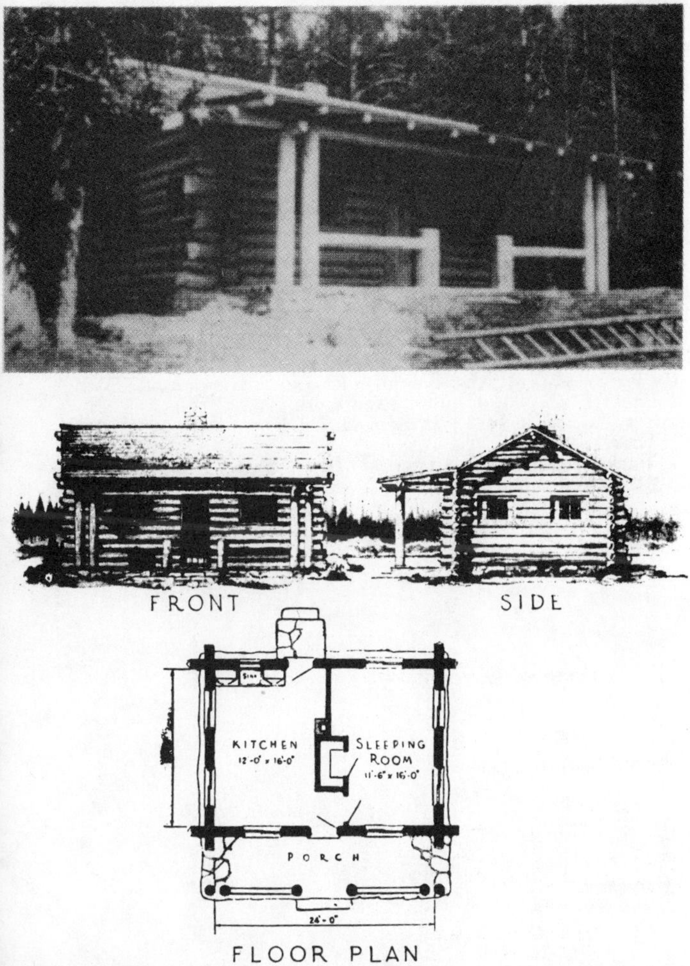

FIGURE 51.—U. S. Forest Service two-room fireguard cabin adaptable for summer residence use.

shows the layout of a United States Forest Service two-room guard cabin adaptable for summer residence use.

ADDITIONAL INFORMATION

Additional useful information on building log cabins may be obtained from the following publications:

UNITED STATES DEPARTMENT OF AGRICULTURE

FIREPLACES AND CHIMNEYS. Farmers' Bul. 1889, 52 pp., illus. 1941.

PROTECTION OF LOG CABINS, RUSTIC WORK, AND UNSEASONED WOOD FROM INJURIOUS INSECTS. Farmers' Bul. 1582, 20 pp., illus. 1929.

USE OF LOGS AND POLES IN FARM CONSTRUCTION. Farmers' Bul. 1660, 26 pp., illus. 1931.

OTHER SOURCES

LOG BUILDINGS. Wis. Agr. Col. Ext. Stencil Cir. 158, 39 pp., illus. 1940.

LOG CABIN CONSTRUCTION. A. B. BOWMAN. Mich. State Col. Ext. Bul. 222, 54 pp., illus. 1941.

LOG CABINS AND COTTAGES: HOW TO BUILD AND FURNISH THEM. W. A. BRUETTE, ed. 96 pp., illus. New York.

THE REAL LOG CABIN. C. D. ALDRICH. 278 pp., illus. 1934. New York.

SHELTERS, SHACKS, AND SHANTIES. D. C. BEARD. 243 pp., illus. 1932. New York.

A Catalog Of Shorey Publications

SHOREY PUBLICATIONS IN PRINT

1. Glover, S.L. ORIGIN AND OCCURENCE OF GEM STONES IN WASHINGTON. 1949 1.00
2. Mooney, Jas. Ghost Dance Religion: SMOHALLA AND HIS DOCTRINE. 1896 1.50
3. Mooney. Ghost Dance Religion: SHAKERS OF PUGET SOUND. 1896 1.50
5. DICTIONARY OF CHINOOK JARGON. nd 2.00
6. ALASKA RAILROAD TIME TABLES. 1922 .50
7. Wells, E.H. UP AND DOWN THE YUKON. 1900 1.50
8. Williams, J.G. REPORT OF ATTORNEY GENERAL, Territory of Alaska, 1949-51. 1.50
9. Gruening, Ernest. MESSAGE OF THE GOVERNOR OF ALASKA, 1946. 2.00
10. Gruening. THIRD MESSAGE TO THE PEOPLE OF ALASKA, 1946. 1.50
11. Kohlstedt, E.D. A GLIMPSE OF ALASKA. 1930 1.00
12. Gruening. MESSAGE OF THE GOVERNOR OF ALASKA, 1945. 2.50
13. Leehey, M.D. PUBLIC LAND POLICY OF THE U.S. IN ALASKA. 1912 1.50
14. NOME TELEPHONE DIRECTORY. 1905 3.00
15. ESTABLISHMENT OF MT. MCKINLEY NATIONAL PARK. 1916 1.50
16. Hooper, Capt. C.L. CRUISE OF THE U.S. REVENUE STEAMER CORWIN IN ARCTIC OCEAN, November 1, 1880. 7.50
17. Gruening. MESSAGE TO THE PEOPLE OF ALASKA. 1945 2.00
18. Gruening. MESSAGES OF THE GOVERNOR OF ALASKA. 1949 2.50
19. Davis, Geo. T. METLAKAHTLA. 1904 3.50
20. WORK OF THE BUREAU OF EDUCATION FOR NATIVES OF ALASKA. 1918 5.00
21. MacDowell, L.W. ALASKA INDIAN BASKETRY. 1.00
22. Eells, Rev. Myron. THE TWANA, CHEMAKUM & KLALLAM INDIANS OF WASHINGTON TERRITORY. 1887 4.00
23. Grinnell, Joseph. GOLD HUNTING IN ALASKA. 4.00
24. Swan, James G. INDIANS OF CAPE FLATTERY. 7.50
25. CUSTER'S LAST BATTLE. 1892 2.00
26. Gruening, Sen. INDEPENDENCE DAY ADDRESS. 1959 1.75
27. Cadell, H.M. THE KLONDIKE AND YUKON GOLDFIELD IN 1913. 1914 2.50
28. Shiels, Archie W. EARLY VOYAGES OF THE PACIFIC. 1930 3.00
29. Fickett, E.D. METEOROLOGY. From Explorations in Alaska. 1900
30. Richardson, W.P. & others. YUKON RIVER EXPLORING EXPEDITION. Expl. in Alaska. 1900 2.00
31. Ray, P.H., W.P. Richardson. RELIEF OF THE DESTITUTE IN THE GOLDFIELDS. 1900 1.75
32. Shiels, Archie W. LITTLE JOURNEYS INTO THE HISTORY OF RUSSIAN AMERICA AND PURCHASE OF ALASKA. 1949 7.50
33. Shiels. STORY OF TWO DREAMS. 1957 4.00
34. Swan, J.G. HAIDAH INDIANS OF QUEEN CHARLOTTE'S ISLAND, B.C. 1874 5.00
35. Butler, Gen. B.V. & the Marquis of Lorne. THE BERING SEA CONTROVERSY. 1892 1.50
36. Crawford, Lewis F. THE MEDORA-DEADWOOD STAGE LINE. 1925 1.50
37. Anderson, Eskil. ASBESTOS AND JADE OCCURENCES IN KOBUK RIVER REGION, Alaska. 1945 2.00
38. Stewart, B.D. PROSPECTING IN ALASKA. 1949 1.50
39. Geoghegan, Richard H. THE ALEUT LANGUAGE. 5.00
40. SEATTLE'S FIRST BUSINESS DIRECTORY. 1876 7.50
41. THE WASHBURN YELLOWSTONE EXPEDITION. 1871 1.25
42. Subreports from EXPLORATIONS IN ALASKA. 7 reports: The Tanana, Chickaloon, Sushitna, etc. 1900 2.00
43. Glenn, Capt. E.F. TANANA RIVER EXPLORING EXPEDITION. 1900 2.50
44. Glenn. COOK'S INLET EXPLORING EXPEDITION. 1900 1.75
45. WILKESON'S NOTES ON PUGET SOUND (1870?) 2.50
46. (Perry). JOURNAL OF A VOYAGE TO THE ARCTIC REGIONS IN H.M.S. ALEXANDER, 1818. (1819?) 6.00
47. BIENNIAL MESSAGE OF WM. M. BUNN, Governor of Idaho. 1884 2.00
48. MACKENZIE'S ROCK. Exploration of Sir Alexander Mackenzie. 1905 2.50
49. Williams, L.R. OUR PACIFIC COUNTY (Wash) 5.00
50. AARON LADNER LINDSLEY..Founder of Alaska Missions. 1.00
51. Evans, Elwood. PUGET SOUND: Its Past, Present and Future. 1869 2.50
52. Campbell, Robt. TWO JOURNALS, 1808-1853. 15.00
53. Bernhardi, Madame Charlotte. MEMOIR OF THE CELEBRATED ADM. JOHN de KRUSENSTERN. 1856 5.00
54. McWhorter, Lucullus V. CRIME AGAINST THE YAKIMAS. 1931 4.50
55. Lupton, Chas T. OIL AND GAS IN THE OLYMPIC PENINSULA. 1913 4.00
56. Seward, Wm. H. THE ADMISSION OF KANSAS. A Speech. 1860 1.50
57. Stevens, Issac I. PROCLAMATION OF MARTIAL LAW. 1856 .50
58. THE GLACIER - Tlinkit Training Academy, Vol. II. 1887 .75
59. Hills, Rev. E.P. REV. AARON L. LINDSLEY. .50
60. SWANTON'S HANDBOOK OF NORTH AMERICAN INDIANS, Pt. C: Indians of Alaska & Canada. 1952 5.00
61. Ward, D.B. ACROSS THE PLAINS IN 1853. 1911 3.50
62. MacDowell, Lloyd W. ALASKA TOTEM POLES. 1.25
63. MacDowell. A TRIP TO WONDERFUL ALASKA. 2.00
64. MacDowell. ALASKA GLACIERS & ICE FIELDS. 1.25
65. Shaw, Geo. C. THE CHINOOK JARGON AND HOW TO USE IT. 1909 4.00

66. PUGET SOUND COOPERATIVE COLONY. 1890's 2.00
67. THE HEROES OF BATTLE ROCK, or The Miners' Reward. 1904 1.50
68. Lampman, Ben Hur. CENTRALIA TRAGEDY AND TRIAL. 1920 4.00
69. SEAL FISHERY IN THE NORTH PACIFIC. (Correspondence with Russia.) 1895 7.50
70. Coolican, J.C. (ed.) PORT ANGELES. 1898 4.00
71. Jordan, D.S. TROUT AND SALMON OF THE PACIFIC COAST. 1906 1.25
72. COOPERATIVE PLAN FOR SECURING HOMES AT PORT ANGELES, Wash. 1893 1.00
73. Hidden, Maria L.T. OREGON PIONEERS. 1910 1.00
74. Stevens, Hazard. FIRST ASCENT OF TAKHOMA. Ext., 1876 1.50
75. DEMOCRATIC AND REPUBLICAN Ticket For Washington State and King County. 1889 .50
76. THIRD ANNUAL COMMENCEMENT OF SEATTLE HIGH SCHOOL. 1888 .35
77. ASSASSINATION OF PRESIDENT GARFIELD. Newspaper Extra. 1881? .50
78. PUGET SOUND CATECHISM. 1889 1.00
79. SEATTLE, KING COUNTY, WASHINGTON TERRITORY. Statistics & descriptive report. 1884 1.50
80. THE GREAT SEATTLE FIRE OF June 6, 1889. 2.50
81. USGS Bull. 442-D. MINING IN THE CHITINA DIST., etc. 1910 2.50
82. Swanton, John R. INDIAN TRIBES OF THE PACIFIC NORTHWEST. Ext., 1952 6.00
83. Sayre, Alex N. PUGET SOUND. Poem 1883 1.50
84. Bell, W.S. OLD FORT BENTON. 1909 2.50
85. Meany, Edmund S. INDIAN GEOGRAPHIC NAMES OF WASHINGTON. 1908 1.00
86. Swanton, J.R. INDIAN TRIBES OF AMERICAN SOUTHWEST. 1952 6.00
87. Swanton. INDIANS OF MEXICO, CENTRAL AMERICA AND WEST INDIES. 1952 2.50
88. USGS Bull. 836. SELECTED LIST OF USGS PUBLICATIONS ON ALASKA. 1930 1.00
89. Lewis, Wm. S. EARLY DAYS IN THE BIG BEND COUNTRY. 1926 3.50
90. Stoney, Geo. M. EXPLORATIONS IN ALASKA. 1899, Ext. from US Naval Inst. 6.00
91. Lindsley, A.L. SKETCHES OF AN EXCURSION TO SOUTHERN ALASKA. 1881 5.00
92. Collins, H.B. ARCHEOLOGY OF THE BERING SEA REGION. Ext., 1933 1.50
93. Ray and Murdoch. A VOCABULARY OF THE ESKIMOS OF POINT BARROW. 1885 1.25
94. Andrews, C.L. STORY OF SITKA. 1922 5.00
95. Matthews, Washington. NAVAJO WEAVERS. 1884 2.50
96. Dawson, Geo. M. THE HAIDAS. 1882 1.00
97. McClintock, Walter. FOUR DAYS IN A MEDICINE LODGE. 1900 1.00
98. Barbeau, Marius. MODERN GROWTH OF THE TOTEM POLES ON THE NORTHWEST COAST. 1939 1.00
99. SCRIPTURE SELECTIONS AND HYMNS IN HIDATSA or GROS VENTURE LANGUAGE. 1906 1.50
100. Baker, Marcus. GEOGRAPHIC DICTIONARY OF ALASKA. 1906 3.00
101. Lee, Charles A. ALEUTIAN INDIAN AND ENGLISH DICTIONARY. 1896 1.00
102. Shaw, Geo. C. VANCOUVER'S DISCOVERY OF PUGET SOUND IN 1792. 1933 1.00
103. Seward, Wm. H. SPEECHES in Alaska, Vancouver, and Oregon, 1869. 2.50
104. Rees, J.E. IDAHO: Chronology, Nomenclature, Biblio. 1918 5.00
105. Smith, W.C. THE EVERETT MASSACRE. 1916 8.00
106. PLACER MINING. 1897 4.50
107. Ross, Clyde P. THE VALDEZ CREEK MINING DISTRICT IN ALASKA. 1933 5.00
108. Stevens, Isaac I. TREATY BETWEEN THE U.S. AND THE DWAMISH, SUQUAMISH, & OTHER ALLIED AND SUBORDINATE TRIBES OF INDIANS IN WASHINGTON TERRITORY. 1855 2.00
109. Stevens, I.I. TREATY BETWEEN THE U.S. & THE NISQUALLY & OTHER BANDS OF INDIANS. 1855 2.00
110. Stevens. TREATY BETWEEN THE U.S. & THE YAKIMA NATION OF INDIANS. 1855 2.00
111. Stevens. TREATY BETWEEN THE U.S. OF A. & THE MAKAH TRIBE OF INDIANS. 1855 2.00
112. Stevens. TREATY BETWEEN THE U.S. AND THE INDIANS OF THE WILLAMETTE VALLEY. 1855 2.00
113. Densmore, F. STUDY OF INDIAN MUSIC. 1941 1.50
114. Harriman, Job. THE CLASS WAR IN IDAHO. 3.00
115. Ranck, G. PICTURES FROM NORTHWEST HISTORY. 2.00
116. Whitaker, R. OVERLAND TO OREGON. 1906 4.00
117. Krieger, H. INDIAN VILLAGES OF S.E. ALASKA. 2.50
118. Gatschet, A. THE KLAMATH INDIANS OF SOUTHWEST OREGON. 1890 4.50
119. Crane, W. TOTEM TALES. 1952 4.00
120. Bate, J. THE SECOND BOOKE TEACHING MOST PLAINLY AND COMPOSING OF ALL MANNER OF FIREWORKS FOR TRYUMPH AND RECREATION. 1635 3.00
121. James, G. PRACTICAL BASKET MAKING. 1917 5.00
122. Whorf, B. MAYA HIEROGLYPHS. 1941 1.25
123. Dall, Wm. MASKS, LABRETS, AND CERTAIN ABORIGINAL CUSTOMS. 1884 8.50
124. McCurdy, J. CAPE FLATTERY AND ITS LIGHT. 1.00
125. Wilhelm, Homer. THE SAN JUAN ISLANDS. 1901 5.00
126. Gaerity, Jack. BREAD AND ROSES FROM STONE. 4.00
127. Smith, W.C. WAS IT MURDER? (Centralia) 1922 3.50
128. HISTORY OF OWYHEE COUNTY. (Idaho) 8.50
129. Evans, Elwood. WASHINGTON TERRITORY. 1877 3.00
130. Delaney, Matilda Sagar. THE WHITMAN MASSACRE. 1920 3.00
131. Hubback, T.W. TEN THOUSAND MILES TO ALASKA FOR MOOSE AND SHEEP. 1921 4.00
132. Wilson, Katherine. COPPER-TINTS. 1923 3.00
133. Coleman, Edmund T. FIRST ASCENT OF MT. BAKER. Ext. from Harper's Monthly. 1869 2.50
134. Hodge, L.K. (ed.) MINING IN THE PACIFIC NORTHWEST. 1897 30.00
134a. Hodge. MINING IN WESTERN WASHINGTON. Ext from #134 10.00

134b. Hodge. *MINING IN CENTRAL & EASTERN WASHINGTON.* Ext. from #134 10.00
134c. Hodge. *MINING IN SOUTHERN BRITISH COLUMBIA.* Ext. from #134 10.00
135. Stallard, Bruce. *ARCHEOLOGY IN WASHINGTON.* 1958 3.00
136. Lockley, Fred. *ALASKA'S FIRST FREE MAIL DELIVERY IN 1900.* 1.25
137. Strange. *JAMES STRANGE'S JOURNAL AND NARRATIVE OF THE COMMERCIAL EXPEDITION FROM BOMBAY TO THE NORTHWEST COAST OF AMERICA.* 1928 5.00
138. Hathaway, Ella C. *BATTLE OF THE BIG HOLE.* 2.00
139. Haller, Granville. *SAN JUAN & SECESSION.* 1.50
140. *THE KLONDIKE NEWS.* Vol. 1 #1, Dawson Newspaper. 1898 (Alaska's rarest newspaper) 10.00
141. *JOURNAL OF MEDOREM CRAWFORD.* 1897 2.00
142. *OCOSTA!* The Ocean Terminus of the Northern Pacific R.R. and Coast City of Washington. 2.50
143. Sutherland, T.A. *HOWARD'S CAMPAIGN AGAINST THE NEZ PERCE INDIANS, 1877.* 1878 3.50
144. Matthews, Mathew. *THE CATLIN COLLECTION OF INDIAN PAINTINGS.* 1890 3.00
145. Lockley, Fred. *TO OREGON BY OX TEAM IN '47.* 1.25
146. Lockley. *VIGILANTE DAYS IN VIRGINIA CITY.* 1.50
147. Judson, Katherine. *MYTHS AND LEGENDS OF THE PACIFIC NORTHWEST.* 1910 7.50
148. Whiting, Dr. F.B. *GRIT, GRIEF AND GOLD.* 10.00
149. Amundsen, Capt. Roald. *TO THE NORTH MAGNETIC POLE AND THROUGH THE NORTHWEST PASSAGE.* 3.00
150. Butte Businessmen's Assn. *BUTTE, MONTANA.* 2.00
151. Immigration Aid Society of Washington Terr. *NORTHWESTERN WASHINGTON.* 1880 3.50
152. Denig, Edwin T. *INDIAN TRIBES OF THE UPPER MISSOURI.* 1930 10.00
153. Buckley, Rev. J.M. *TWO WEEKS IN THE YOSEMITE AND VICINITY.* 1888 2.50
154. *SPEECH OF THE HON. R.C. WINTHROP OF MASS. ON THE PRESIDENT'S MESSAGE* (on admission of California) 1850 2.00
155. Marshall, Martha. *A PRONOUNCING DICTIONARY OF CALIFORNIA NAMES IN ENGLISH AND SPANISH.* 2.50
156. Mercer, A.S. *BIG HORN COUNTY, WYOMING.* 7.50
157. Moorehead, Warren K. *PREHISTORIC RELICS.* 7.50
158. Merrill, Geo. P. *NOTES ON THE GEOLOGY AND NATURAL HISTORY OF THE PENINSULA OF LOWER CALIFORNIA.* 1895 2.00
159. Georgeson, C.C. *REINDEER & CARIBOU.* 1904 1.50
160. Hewett, Edgar L. *A GENERAL VIEW OF THE ARCHEOLOGY OF THE PUEBLO REGION.* 1904 2.00
161. Bennett, Wm. P. *THE FIRST BABY IN CAMP.* 3.50
162. Coffman, Noah B. *OLD LEWIS COUNTY, Oregon Territory.* 1926 3.50
163. Bebbe, Mrs. Iola. *THE TRUE LIFE STORY OF SWIFTWATER BILL GATES.* 1908 7.50
164. Costello. *THE SIWASH - THEIR LIFE, LEGENDS AND TALES.* 1895 10.00
165. Murie, Olaus J. *ALASKA-YUKON CARIBOU.* 10.00
166. Meeker, Ezra. *STORY OF THE LOST TRAIL TO OREGON, NO. 2.* 1916 2.50
167. Hanna, Rev. J.A. *DR. WHITMAN AND HIS RIDE TO SAVE OREGON.* 1903 1.25
168. Hadwen, Seymore & Laurence J. Palmer. *REINDEER IN ALASKA.* 1922 5.00
169. Geer, T.T. *THE ROMANCE OF ASTORIA.* 1911 1.75
170. Williams, Lewis. *CHINOOK BY THE SEA.* 7.50
171. *SAN FERNANDO VALLEY.* 1938 2.00
172. *PASADENA.* 1938 2.00
173. Ingalls, Maj. J.W. *HISTORY OF WASHOE COUNTY, NEVADA.* 1913 5.00
174. Reid, J.T. & J.R. Hunter. *HUMBOLT COUNTY.* 1913 4.50
175. Aston. *ESMERALDA COUNTY.* 1913 4.50
176. London, Jack. *THE GOLD HUNTERS OF THE NORTH.* Ext., 1903. 1.25
177. Wartman-Arland, Flora E. *THE STORY OF MONTESANO.* 1933 4.00
178. London. *THE ECONOMICS OF THE KLONDIKE.* 1.50
179. Bishop, Robert Sr. *LAND IN THE SKY TOTEM.* 1.50
180. Webb, John S. & Ed S. Curtis. *THE RIVER TRIP TO THE KLONDIKE AND THE RUSH TO THE KLONDIKE OVER THE MOUNTAIN PASS.* 1898 2.50
181. Librn. of Bellingham Public Libr. *HISTORY OF BELLINGHAM.* 1926 6.00
182. *VOYAGE OF ALEXANDER MACKENZIE.* 2.50
183. Chaplin, Ralph. *CENTRALIA CONSPIRACY.* 4.50
184. Haswell, Robert. *ROBERT HASWELL'S JOURNALS.* 1788-89. 4.00
185. Minto, John. *RHYME OF EARLY LIFE IN OREGON.* (1915?) 4.00
186. Collins, Henry B., A.H. Clarke & E.H. Walker. *THE ALEUTIAN ISLANDS: Their People & Natural History.* 1945 7.50
187. Simpson, Brevet Brig. Gen. J.H. *CORONADO'S MARCH IN SEARCH OF THE "SEVEN CITIES OF CIBOLA."* 1871 2.50
188. Riddle, Geo. W. *EARLY DAYS IN OREGON.* 5.00
189. *VILHJALMUR STEFANSON.* 1925 2.50
190. Lockley, Fred. *ACROSS THE PLAINS BY PRAIRIE SCHOONER.* 1.50
191. MacArthur, Walter. *LAST DAYS OF SAIL ON THE WEST COAST.* 1929 7.50
192. Stuck, Hudson. *T E ALASKAN MISSIONS OF THE EPISCOPAL CHURCH.* 1920 10.00
193. Steffa, Don. *TALES OF NOTED FRONTIER CHARACTERS, SOAPY SMITH.* 1908 1.75.
194. Meeker, E. *WASGUBGTIB TERRUTIRT,* k870 3.00
195. Denny, Arthur A. *PIONEER DAYS ON PUGET SOUND.* 1888 5.00
196. Rowan, James. *THE I.W.W. IN THE LUMBER INDUSTRY.* 4.00
199. Sayre, J. Willis. *THE EARLY WATERFRONT OF SEATTLE, 1937.* 2.50

204. Pelly, T.M. DR. MINOR - A SKETCH OF HIS BACKGROUND AND LIFE. 1933 7.50
209. Flandrau, Grace. FRONTIER DAYS ALONG THE UPPER MISSOURI. 3.00
210. Flandrau. THE LEWIS AND CLARK EXPEDITION. 5.00
211. Flandrau. A GLANCE AT THE LEWIS AND CLARK EXPEDITION. 3.00
213. Flandrau. THE VERENDRYE OVERLAND QUEST OF THE PACIFIC. 3.50
216. Harper, Frank B. FORT UNION AND ITS NEIGHBORS ON THE UPPER MISSOURI. 2.50
220. Burdick, Usher L. MARQUIS de MORES AT WAR IN THE BADLANDS. 1929 2.50
227. General Strike Committee. THE SEATTLE GENERAL STRIKE. 1919 5.00
231. Sayre, J.W. THE ROMANCE OF SECOND AVENUE. 1.50
235. DEDICATION AND OPENING OF THE NEW CASCADE TUNNEL. 1929 2.00
239. Van Olinda, O.S. HISTORY OF VASHON - MAURY ISLAND. 1935 6.00
241. Hanford, C.H. SAN JUAN DISPUTE. 1900 2.00
246. Walgamott, C.S. REMINISCENCES OF EARLY DAYS. 1926 7.50
250. Wardner, Jim. JIM WARDNER, OF WARDNER IDAHO. 1900 8.00
251. Brown, Col. W.C. THE SHEEPEATER CAMPAIGN, Idaho - 1879. 1926 2.50
259. Hornaday, Wm. T. EXTERMINATION OF THE AMERICAN BISON. 1887? 10.00
264. Slauson, Morda C. ONE HUNDRED YEARS ON THE CEDAR. 10.00
265. Jones, S.C. & M.F. Casady. FROM CABIN TO CUPOLA. County Courthouses in Washington. 10.00
267. Andrews, Clarence L. WRANGELL AND THE GOLD OF THE CASSIAR. 1937 3.50
268. Andrews, Clarence L. THE PIONEERS AND THE NUGGETS OF VERSE. 1937 3.00
280. PORT TOWNSEND. 1890 5.00
281. Gilbert, Kenneth. ALASKAN POKER STORIES. 1958 2.50
282. Moore, James Bernard. SKAGWAY IN DAYS PRIMEVAL. 4.00
283. Chase, Cora G. UNTO THE LEAST - A Biographic Sketch of Mother Ryther. 1972 6.00
284. Blankenship, Geo. E. EARLY HISTORY OF THURSTON COUNTY. 1914 20.00
285. Fish, Harriet U. PAST AT PRESENT In Issaquah, Wn. 1967 12.50
286. Allen, Edward W. THE ROLLICKING PACIFIC; A Selection of Poems. 1972 3.00
287. Allen. DANCING TALES AND OTHER FISHY JINGLES. 1951 4.00
289. Chin, Doug & Art. UPHILL - The Settlement & Diffusion of the Chinese in Seattle. 6.00
292. Harpham, Josephine E. DOORWAYS INTO HISTORY - Early Houses & Public Buildings of Oregon. 5.00

SJU 1. Farlow, Dr. W.G. SOME EDIBLE AND POISONOUS FUNGI. 1897 2.50
SJU 2. Gurdji, V. ORIENTAL RUG WEAVING. 1901 5.00
SJU 3. D'Avenes, E., Prisse & Dr. J.C. Ewart. EGYPTIAN AND ARABIAN HORSES AND ORIGIN OF HORSES AND PONIES. 1904 2.50
SJU 4. Freshfield, Douglas W. ON MOUNTAINS AND MANKIND. 1904 1.50
SJU 5. Davenport, Cyril. CAMEOS. 1904 1.50
SJU 6. Maire, Albert. MATERIALS USED TO WRITE UPON BEFORE THE INVENTION OF PRINTING. 1.50
SJU 7. Liberty, Arthur L. PEWTER AND THE REVIVAL OF ITS USE. 1904 1.50
SJU 8. Hammell, Wm. PINE NEEDLE BASKETRY. 1.50
SJU 9. Frey, Clark, Vietch. HOME TANNING. 1936 1.50
SJU 10. Browning, Frank. STEAM PLANT ERRORS 2.00
SJU 18. Burdick, Arthur J. THE PROSPECTOR'S MANUAL. 1905 7.50
SJU 19. WOODS DUAL POWER. (1902?) 1.50
SJU 20. Hayward, Charles B. DIRIGIBLE BALLOONS. 1921 5.00
SJU 21. Osborn, Henry F. THE ELEPHANTS AND MASTEDONS ARRIVE IN AMERICA. 1.50
SJU 22. Todd, Mattie Phipps. HAND LOOM WEAVING. 1902 7.00
SJU 23. Wilbur, C. Martin. HISTORY OF THE CROSSBOW. 1936 1.50
SJU 24. Alexander, A.S. HORSE SECRETS. 1913 2.50
SJU 26. Clute, Willard Nelson. THE FERN - COLLECTOR'S GUIDE. 1901 3.00
SJU 27. Meeker, E. HOP CULTURE IN THE U.S. 1883 10.00
SJU 28. Austin-Walker Sales Co. APACHE BEADWORK - Instructions and Designs 2.50
SJU 29. Holmstrom, J.G. SCIENTIFIC HORSE, MULE & OX SHOEING. 1902 4.00
SJU 30. Fickes, Clyde P. & W. Ellis Groben. BUILDING WITH LOGS. 1945 2.50

SJI 1. Krause, F. SLING CONTRIVANCES. 1904 2.00
SJI 2. Murdoch, John. A STUDY OF THE ESKIMO BOWS IN THE U.S. NATIONAL MUSEUM. 1884 1.50
SJI 3. Mason, Otis T. THROWING STICKS IN THE NATIONAL MUSEUM. 1884 2.50
SJI 4. CHIEF JOSEPH'S OWN STORY. 1879 2.50
SJI 5. Mason, Otis. BASKET WORK OF THE ABORIGINES. 1884 5.00
SJI 6. Llwd, Rev. Dr. J.P.D. THE MESSAGE OF AN INDIAN RELIC: Seattle's Totem Pole. 2.00
SJI 7. Mason, O. TRAPS OF THE AMERICAN INDIAN. 1.25
SJI 8. Meeker, Louis L. OGALALA GAMES. 1901 1.50
SJI 9. Collins, Henry B. PREHISTORIC ART OF THE ALASKAN ESKIMO. 1929 3.50
SJI 10. Eells, Rev. M. JUSTICE TO THE INDIAN. 1.00
SJI 11. Willoughby, C. INDIANS OF THE QUINAIELT AGENCY, Washington Territory. 1886 1.50
SJI 12. Leon & Holmes. STUDIES ON THE ARCHAEOLOGY OF MICHOACAN MEXICO...1886 2.00
SJI 13. Beckwith, Paul. NOTES ON CUSTOMS OF THE DAKOTAHS. 1886 1.25
SJI 14. Eells, Rev. M. THE STONE AGE. 1886 1.50
SJI 15. Allen, Lt. Henry T. ATNATANAS NATIVES OF COPPER RIVER, ALASKA. 1886 1.00
SJI 16. Yates, Dr. L.G. CHARM STONES. 1886 1.50

SJI 20.	*Gibbs, Geo., Dr. Wm. F. Tolmie, & Father G. Mengarini. TRIBES OF WESTERN WASHINGTON AND NORTHWESTERN OREGON.*	*20.00*
SJI 21.	*Gibbs, Geo & others. COMPARATIVE VOCABULARIES OF THE TRIBES OF WESTERN WASHINGTON AND NORTHWESTERN OREGON. part of SJI 20.*	*5.00*
SJI 22.	*Gibbs, Geo. A DICTIONARY OF THE NISQUALLY INDIAN LANGUAGE. part of SJI 20.*	*10. 0*
SJI 23.	*Boas, Franz. INTRODUCTION TO HANDBOOK OF AMERICAN INDIAN LANGUAGES. Ext. 1911*	*5.00*
SJI 24.	*Swanton, John R. THE TLINGIT INDIAN LANGUAGE. From Hndbk. of Amer. Indian Lang.*	*3.00*
SJI 25.	*Swanton. THE HAIDA INDIAN LANGUAGE.*	*5.00*
SJI 26.	*Boas. THE TSIMSHIAN INDIAN LANGUAGE.*	*8.00*
SJI 27.	*Boas. THE KWAKIUTL INDIAN LANGUAGE.*	*7.50*
SJI 28.	*Boas. THE CHINOOK INDIAN LANGUAGE.*	*5.00*
SJI 29.	*Boas & Swanton. THE SIOUAN, Dakota INDIAN LANGUAGE.*	*6.00*
SJI 30.	*Talbitzer, Wm. THE ESKIMO LANGUAGE.*	*6.50*
SJI 31.	*Ewers, John C. EARLY WHITE INFLUENCE UPON PLAINS INDIAN PAINTING. 1957*	*2.00*
SJI 32.	*Fenton, Wm. N. CONTRACTS BETWEEN IROQUOIS HERBALISM AND COLONIAL MEDICINE. 1941*	*2.00*
SJI 38.	*Remington, Fred. ARTIST WANDERINGS AMONG THE CHEYENNES. 1889*	*1.50*
SJI 41.	*Bagley, Clarence. INDIAN MYTHS OF THE NORTHWEST. 1930*	*8.50*
SJI 43.	*Mason, O. ABORIGINAL SKIN DRESSING.*	*5.00*
SJI 46.	*Thorne, J. Frederic. IN THE TIME THAT WAS. 1909*	*2.50*
SJI 49.	*DeSmet, Rev. P.J. NEW INDIAN SKETCHES.*	*8.50*
SJI 51.	*Mason, O. MAN'S KNIFE AMONG THE NORTH AMERICAN INDIANS. 1899*	*2.00*
SJI 53.	*Holmes, William H. POTTERY OF THE ANCIENT PUEBLOS. 1886*	*6.00*
SJI 55.	*Dorsey, James O. A STUDY OF SIOUAN CULTS 1894*	*10.00*
SJI 57.	*Mason, O. POINTED BARK CANOES OF THE KUTENAI AND AMUR. 1901*	*2.00*
SJI 58.	*Mason, O. ABORIGINAL AMERICAN HARPOONS. 1902*	*7.50*
SJI 59.	*Hough, Walter. THE LAMP OF THE ESKIMO. 1898*	*5.00*
SJI 61.	*Drucker, Philip. ARCHAEOLOGICAL SURVEY CN THE NORTHERN NORTHWEST COAST. 1943.*	*7.00*
SJI 62.	*INDIANS IN WASHINGTON.*	*3.00*

ISBN # 0-8466-6030-X